CRC SERIES IN RADIOTRACERS IN BIOLOGY AND MEDICINE

Editor-in-Chief

Lelio G. Colombetti, Sc.D.
Loyola University
Stritch School of Medicine
Maywood, Illinois

STUDIES OF CELLULAR
FUNCTION USING
RADIOTRACERS
Mervyn W. Billinghurst, Ph.D.
Radiopharmacy
Health Sciences Center
Winnipeg, Manitoba, Canada

RECEPTOR-BINDING
RADIOTRACERS
William C. Eckelman, Ph.D.
Department of Radiology
George Washington University School
of Medicine
Washington, D.C.

GENERAL PROCESSES OF
RADIOTRACER LOCALIZATION
Leopold J. Anghileri, D.Sc.
Laboratory of Biophysics
University of Nancy
Nancy, France

BIOLOGIC APPLICATIONS OF
RADIOTRACERS
Howard J. Glenn, Ph.D.
University of Texas System Cancer
Center
M.D. Anderson Hospital and Tumor
Institute
Houston, Texas

RADIATION BIOLOGY
Donald Pizzarello, Ph.D.
Department of Radiology
New York University Medical Center
New York, New York

BIOLOGICAL TRANSPORT OF
RADIOTRACERS
Lelio G. Colombetti, Sc.D.
Loyola University
Stritch School of Medicine
Maywood, Illinois

RADIOTRACERS
FOR MEDICAL
APPLICATIONS
Garimella V. S. Rayudu, Ph.D
Nuclear Medicine Department
Rush University Medical Center
Presbyterian - St. Luke's Hospital
Chicago, Illinois

BASIC PHYSICS
W. Earl Barnes, Ph.D.
Nuclear Medicine Service
Edward Hines, Jr., Hospital
Hines, Illinois

RADIOBIOASSAYS
Fuad S. Ashkar, M.D.
Radioassay Laboratory
Jackson Memorial Medical Center
University of Miami School of Medicine
Miami, Florida

Basic Physics of Radiotracers

Volume II

Editor

W. Earl Barnes, Ph.D.
Physicist
Nuclear Medicine Services
Edward Hines, Jr., Veterans Administration Hospital
Hines, Illinois

Editor-in-Chief
CRC Series in Radiotracers in Biology and Medicine

Lelio G. Colombetti, Sc.D.
Loyola University
Stritch School of Medicine
Maywood, Illinois

CRC Press, Inc.
Boca Raton, Florida

Library of Congress Cataloging in Publication Data
Main entry under title:

Basic physics of radiotracers.

  (CRC series in radiotracers in biology and medicine)
  Bibliography: p.
  Includes index.
  1. Nuclear physics. 2. Radioactive decay.
I. Barnes, W. Earl. II. Series.
QC776.B3   1983   539.7   82-14723
ISBN 0-8493-6001-3 (v. 1)
ISBN 0-8493-6002-1 (v. 2)

Direct all inquiries to CRC Press, Inc., 2000 Corporate Blvd., N.W., Boca Raton, Florida, 33431.

© 1983 by CRC Press, Inc.

International Standard Book Number 0-8493-6001-3 (v. 1)
International Standard Book Number 0-8493-6002-1 (v. 2)

Library of Congress Card Number 82-14723
Printed in the United States

AAD-5851

FOREWORD

This series of books on Radiotracers in Biology and Medicine is on the one hand an unbelievably expansive enterprise and on the other hand, a most noble one as well. Tools to probe biology have developed at an accelerating rate. Hevesy pioneered the application of radioisotopes to the study of chemical processes, and since that time, radioisotopic methodology has probably contributed as much as any other methodology to the analysis of the fine structure of biologic systems. Radioisotopic methodologies represent powerful tools for the determination of virtually any process of biologic interest. It should not be surprising, therefore, that any effort to encompass all aspects of radiotracer methodology is both desirable in the extreme and doomed to at least some degree of inherent failure. The current series is assuredly a success relative to the breadth of topics which range from in depth treatise of fundamental science or abstract concepts to detailed and specific applications, such as those in medicine or even to the extreme of the methodology for sacrifice of animals as part of a radiotracer distribution study. The list of contributors is as impressive as is the task, so that one can be optimistic that the endeavor is likely to be as successful as efforts of this type can be expected to be. The prospects are further enhanced by the unbounded energy of the coordinating editor. The profligate expansion of application of radioisotopic methods relate to their inherent and exquisite sensitivity, ease of quantitation, specificity, and comparative simplicity, especially with modern instrumentation and reagents, both of which are now readily and universally available. It is now possible to make biological measurements which were otherwise difficult or impossible. These measurements allow us to begin to understand processes in depth in their unaltered state so that radioisotope methodology has proved to be a powerful probe for insight into the function and perturbations of the fine structure of biologic systems. Radioisotopic methodology has provided virtually all of the information now known about the physiology and pathophysiology of several organ systems and has been used abundantly for the development of information on every organ system and kinetic pathway in the plant and animal kingdoms. We all instinctively turn to the thyroid gland and its homeostatic interrelationships as an example, and an early one at that, of the use of radioactive tracers to elaborate normal and abnormal physiology and biochemistry, but this is but one of many suitable examples. Nor is the thyroid unique in the appreciation that a very major and important residua of diagnostic and therapeutic methods of clinical importance result from an even larger number of procedures used earlier for investigative purposes and, in some instances, procedures used earlier for investigative purposes and, in some instances, advocated for clinical use. The very ease and power of radioisotopic methodology tempts one to use these techniques without sufficient knowledge, preparation or care and with the potential for resulting disastrous misinformation. There are notable research and clinical illustrations of this problem, which serve to emphasize the importance of texts such as these to which one can turn for guidance in the proper use of these powerful methods. Radioisotopic methodology has already demonstrated its potential for opening new vistas in science and medicine. This series of texts, extensive though they be, yet must be incomplete in some respects. Multiple authorship always entails the danger of nonuniformity of quality, but the quality of authorship herein assembled makes this likely to be minimal. In any event, this series undoubtedly will serve an important role in the continued application of radioisotopic methodology to the exciting and unending, yet answerable, questions in science and medicine!

Gerald L. DeNardo, M.D.
Professor of Radiology, Medicine,
Pathology and Veterinary Radiology
University of California, Davis-
Sacramento Medical School
Director, Division of Nuclear Medicine

THE EDITOR-IN-CHIEF

Lelio G. Colombetti, Sc.D., is Professor of Pharmacology at Loyola University Stritch School of Medicine in Maywood, Ill. and a member of the Nuclear Medicine Division Staff at Michael Reese Hospital and Medical Center in Chicago, Ill.

Dr. Colombetti graduated from the Litoral University in his native Argentina with a Doctor in Sciences degree (summa cum laude), and obtained two fellowships for postgraduate studies from the Georgetown University in Washington, D.C., and from the M.I.T. in Cambridge, Mass. He has published more than 150 scientific papers and is the author of several book chapters. He has presented over 300 lectures both at meetings held in the U.S. and abroad. He organized the First International Symposium on Radiopharmacology, held in Innsbruck, Austria, in May 1978. He also organized the Second International Symposium on Radiopharmacology which took place in Chicago in September, 1981, with the active participation of more than 500 scientists, representing over 30 countries. He is a founding member of the International Association of Radiopharmacology, a nonprofit organization, which congregates scientists from many disciplines interested in the biological applications of radiotracers. He was its first President (1979/1981).

Dr. Colombetti is a member of various scientific societies, including the Society of Nuclear Medicine (U.S.) and the Gesellschaft für Nuklearmedizin (Europe), and is an honorary member of the Mexican Society of Nuclear Medicine. He is also a member of the Society of Experimental Medicine and Biology, the Coblenz Society, and the Sigma Xi. He is a member of the editorial boards of the journals *Nuklearmedizin* and *Research in Clinic and Laboratory*.

PREFACE

Few fields of research draw upon a more diverse array of scientific disciplines than does that of radiotracer research in biology and medicine. To master these all would tax the abilities of a Leonardo da Vinci, and as a consequence the researcher becomes expert only in those areas most germane to his work. One discipline frequently receiving short shrift in the education of persons working in the biomedical radiotracer field is the physics of the atom and nucleus as it relates to the nature of radioactive decay and of radiation. This subject is fundamental to all radiotracer experiments and in addition forms the underpinnings of the more applied fields of nuclear instrumentation, radiochemistry, radionuclide production, and radiation dosimetry.

The opportunity to present the physics of radioactive processes in some detail and apart from topics such as instrumentation which conventionally compete with it for space is most welcome. The material is intended to give a fairly complete introduction to radiation physics to those who wish to have more than a descriptive understanding of the subject. Although it is possible to work one's way through much of the subject matter without having had any previous physics background, some prior acquaintance with modern physics is desirable. A familiarity with calculus and differential equations is also assumed.

Volume I begins with a brief description of classical physics, its extension to special relativity and quantum mechanics, and an introduction to basic atomic and nuclear concepts. A thorough discussion of atomic structure follows with emphasis on the theory of the multielectron atom, characteristic X-rays, and the Auger effect. Volume II treats the subjects of nuclear structure, nuclear decay processes, the interaction of radiation with matter, and the mathematics of radioactive decay.

Lelio G. Colombetti, Sc.D.
W. Earl Barnes, Ph.D.

THE EDITOR

W. Earl Barnes, Ph.D., is Physicist at the Veterans Hospital, Hines, Illinois.

Dr. Barnes received the B.A. degree from Harvard University in 1961 and the Ph.D. degree in nuclear physics in 1970 from Oregon State University. In addition to a teaching program in physics, he is engaged in research which applies mathematics and physics to various problems of medicine, having published approximately 40 articles in the areas of mathematical modeling of liver and bone kinetics, functional imaging of the heart and lungs, and radiation dosimetry. Currently Treasurer of the International Association of Radiopharmacology, he has been active in the organization of the First and Second International Symposiums on Radiopharmacology.

Dr. Barnes is a member of Sigma Xi, Sigma Pi Sigma, and the Society of Nuclear Medicine.

CONTRIBUTORS

Harry T. Easterday, Ph.D.
Professor of Physics
Oregon State University
Corvallis, Oregon

B. T. A. McKee, Ph.D.
Associate Professor
Department of Physics
Assistant Professor
Department of Medicine
Queen's University
Kingston, Ontario, Canada

R. R. Sharma, Ph.D.
Professor
Department of Physics
University of Illinois
Chicago, Illinois

Donald A. Walker, Ph.D.
Professor
Department of Physics
State University of New York
New Paltz, New York

TABLE OF CONTENTS

Volume I

Chapter 1
Introductory Concepts 1
W. Earl Barnes

Chapter 2
Aomic Structure 81
R. R. Sharma

Index 201

Volume II

Chapter 1
Nuclear Structure 1
W. Earl Barnes

Chapter 2
Modes of Nuclear Decay 35
Donald A. Walker

Chapter 3
Interaction of Radiation with Matter 79
Harry T. Easterday

Chapter 4
Mathematics of Radioactive Decay 105
B. T. A. McKee

Index 161

Chapter 1

NUCLEAR STRUCTURE

W. Earl Barnes

TABLE OF CONTENTS

I.   Introduction 2

II. Global Properties of the Nucleus 2
     A.   Nuclear Density and Shape..2
     B.   Binding Energy...3
     C.   Angular Momentum 4
     D.   Parity...5
     E.   Magnetic Moments 6
     F.   Electric Moments 7

III. The Two-Nucleon Force 8
     A. $V_{ordinary}$ 9
     B. $V_{exchange}$ 9
     C. V_{tensor} 12
     D. $V_{velocity-dependent}$ 12

IV. Nuclear States 13
     A.   Doubly Magic Nuclides and Their Immediate Neighbors 13
     B.   Nuclides Somewhat Removed from Doubly Magic 22
          1.   Even-Even Nuclides 24
          2.   Odd-Even Nuclides...25
     C.   Nuclides Far Removed from Doubly Magic 26
          1.   Rotational Excitations 27
          2.   Vibrational Excitations 29
          3.   Single-Particle Excitations.....................................29
               a.   Odd-Even Nuclides..31
               b.   Odd-Odd Nuclides 31
               c.   Even-Even Nuclides 34

References 35

I. INTRODUCTION

Whereas there exists a completely satisfactory theory for the behavior of the atomic electrons, the same cannot be said for nucleons. The reasons are that, first of all, the force existing between two nucleons is complicated and incompletely understood. Second, the mathematical problem of dealing with an assembly of A nucleons is inherently difficult, requiring 3A independent variables for the description of nucleon motion and more if spin or other variables are significant. Of course, one may hope that the average effect of internucleon forces may be fairly easily expressed. This is certainly the case for electrons where successful theories are based on a simple average potential produced by the interaction of electrons with the nucleus and with themselves. But unlike electrons, nucleons are not attracted to a single central body, and internucleon forces are strong, short range, and attractive.

Although the prospects are dim of providing a theoretical framework from first principles for the observed characteristics of the nucleus, substantial progress has been made in describing to some degree the properties of nuclei in terms of *nuclear models*. These are highly simplified views of the physics of the nucleus, each of which is typically successful in predicting a limited number of nuclear properties for a particular class of nuclides. In recent years "unified" models have been devised which appear capable of explaining a wide variety of nuclear properties, at least qualitatively.

The number of topics associated with nuclear structure are many. We will limit our discussion to an attempt to understand the properties of the low-lying nuclear states: their energies, angular momenta, parities, and electric and magnetic moments. It is these properties which largely determine the manner in which radionuclides decay.

II. GLOBAL PROPERTIES OF THE NUCLEUS

A brief description of some of the basic properties of nuclei in their ground state (lowest energy state) is presented below.

A. Nuclear Density and Shape

Experiments in which electrons are scattered from nuclei yield information about the density distribution of electric charge, and hence of protons, in the nucleus. Since neutrons are known to be distributed in the same way as protons, the distribution of mass throughout the nuclear volume can be determined. It is found that the nuclear density is approximately constant in the interior of the nucleus, tailing off to zero at the periphery. Furthermore, the nuclear density is approximately constant from nuclide to nuclide, having a value of about 0.2 nucleon masses per fm^3 (fm = 10^{-15}m). Thus the nuclear volume is proportional to the mass of the nucleus or to the number of nucleons A. The nuclear radius R is then proportional to $A^{1/3}$, electron scattering data giving the result:

$$R = 1.07\,A^{1/3}\text{ fm} \tag{1}$$

The independence of nuclear density from mass number indicates that nucleons do not coalesce into each other. This property, which is known as the saturation of nuclear density, has important implications with regard to the nature of the nuclear force.

Thus far the precise shape of the nucleus has not been considered. One of the parameters most sensitive to nuclear shape is the electric quadrupole moment which is discussed in Section II.F. Measurements of nuclear quadrupole moments indicate that nuclei are spherical when their number of protons or neutrons is equal to one of the following numbers: 8, (16), 20, 28, (38), 50, 82, 126. Those numbers not in parentheses are the so-called *magic numbers* at which many nuclear parameters take on anomalous values. Nuclei with Z and N several units removed from magic numbers can be significantly nonspherical. If an ellipsoidal shape is assumed for these deformed nuclei, their quadrupole moments indicate a difference between the major and minor axes of the ellipse which is as much as 30% of the mean axis length.

B. Binding Energy

The binding energy of the nucleus is an especially revealing quantity, providing important information about nuclear forces and structure. Nuclear binding energy is the energy required to completely dissociate the nucleus into free nucleons. Because of the equivalence of energy and mass, binding energy is manifested as a reduction in the mass of the nucleons when assembled. Calling the mass of a free proton m_p and of a free neutron m_n, we have for the binding energy of a nucleus

$$B = (Z\,m_p + N\,m_n - {}^A_Z M)\,c^2$$

where ${}^A_Z M$ is the mass of the nucleus having Z protons and $N = A - Z$ neutrons.

Measurements of nuclear binding energies reveal the following general trends:

1. To an approximation, B is proportional to the number of nucleons A, the proportionality constant being about 8 MeV. This fact suggests that each nucleon interacts with only a limited number of others so that the binding energy per nucleon is independent of the total number of nucleons. Something similar occurs in liquids where each atom is bound to only a few neighbors. A pursuance of this analogy has led to a simple model of the nucleus called the *liquid drop model*. The binding energy of the nucleus is considered to be due to both volume and surface effects. If every nucleon were completely surrounded by neighbors, as it is within the interior of the nucleus, the nuclear binding energy would be strictly proportional to A:

$$B_{volume} = a_1\,A$$

    The magnitude of the proportionality constant a_1 is presumably related to the strength of the internucleon forces and to the maximum number of neighbors with which a nucleon can interact.

2. However, a nucleon at the surface of the nucleus has neighbors on only one side, suggesting that a correction term is needed which is negative and proportional to the number of nucleons at the surface. This number is proportional to the nuclear surface area or to $A^{2/3}$. Thus we have

$$B_{surface} = -a_2\,A^{2/3}$$

3. For high-A nuclides, the binding energy per nucleon is found to fall. An obvious reason for this decrease is the large number of protons in the nucleus which are repelling each other. As the number of protons increases, the effect of Coulomb repulsion, a long-range force, becomes increasingly important relative to the short-range attraction of nuclear forces. The negative energy contribution to B due to Z protons uniformly distributed in a sphere can be shown to be proportional to Z^2 and inversely proportional to the radius or to $A^{1/3}$:

$$B_{coulomb} = -a_3\,\frac{Z^2}{A^{1/3}}$$

4.   Nuclei have larger binding energies when the number of protons is approximately equal to the number of neutrons. To mathematically express this fact, we need a negative binding energy term which increases in magnitude as $|N - Z|$ or $|A - 2Z|$ increases. We use

$$B_{symmetry} = -a_4 \frac{(A/2 - Z)^2}{A}$$

Coulomb repulsion in high-A nuclides tends to increase N at the expense of Z, but the energy effect of this process has already been accounted for in $B_{Coulomb}$.

5.   Nuclei having even numbers of protons and/or neutrons have larger binding energies. This phenomenon is associated with the fact to be discussed below that there is a pairing energy between two like nucleons in the same energy state. This effect is approximately expressed as

$$B_{pairing} = \begin{cases} a_5 \ A^{-3/4} & \text{even Z-even N} \\ 0 & \text{even-odd} \\ -a_5 \ A^{-3/4} & \text{odd Z-odd N} \end{cases}$$

The complete formula for the nuclear binding energy is then

$$B = B_{volume} + B_{surface} + B_{coulomb} + B_{symmetry} + B_{pairing}$$

(2)

Values of the constants (in MeV) giving the best fit to the empirical data are $a_1 = 15.76$, $a_2 = 17.81$, $a_3 = 0.7105$, $a_4 = 94.81$, $a_5 = 34$.

Although Equation 2 fits experimental binding energy data reasonably well over the entire range of stable nuclides, discrepancies occur at certain values of N and Z where measured values of B are larger than expected. These discrepancies arise for N and/or Z equal to one of the magic numbers (2, 8, 20, 50, 82, 126). Magic nuclei and especially "doubly magic" nuclei (both N and Z equal to a magic number) are particularly stable systems. The energy required to remove a neutron, for example, from a nucleus that is magic in N is unusually large while the neutron separation energy for a magic-plus-one-neutron system is unusually small. This situation is reminiscent of the abrupt changes in ionization energy at the shell closures for the atomic electrons.

C. Angular Momentum

As for atomic electrons, each nucleon has an orbital angular momentum l and an intrinsic spin angular momentum s. Associated with l is the quantum number l which is an integer, while the intrinsic spin quantum number s has only the single value one half. Depending on the physical circumstances, the l and s vectors of the nucleons may combine in a variety of ways. Irrespective of these details the total angular momentum J of the nucleus is the sum of these individual angular momenta and is constant in time. The magnitude of J is given by

$$|J| = \sqrt{J(J + 1)} \ \hbar$$

where J is the quantum number associated with J. Commonly, J is referred to as the *spin* of the nucleus, and one must be careful not to confuse this term with the intrinsic spin s of a nucleon.

5

The spin of the nucleus can be observed by measuring the component of J along a particular direction, e.g., the direction of a magnetic field. Customarily, the z-axis is chosen as this direction. Several values are possible for J_z:

$$J_z = m_J \hbar$$

where m_J can take on any of the values

$$m_J = -J, -J + 1, \cdots, J$$

So the number of possible orientations of J relative to the z-axis is $(2J + 1)$.

We recall that if two angular momenta j_1 and j_2 sum to form j

$$j = j_1 + j_2$$

that then the quantum number j associated with j can take on any of the values

$$j = |j_1 - j_2|, |j_1 - j_2| + 1, \cdots, j_1 + j_2$$

where j_1 and j_2 are the quantum numbers associated with j_1 and j_2. These relations can easily be extended to the sum of many angular momenta. Now, in the summation of the l and s vectors of the nucleons to form the total angular momentum J, it can be seen from these addition rules that, because ℓ is integer and s half-integer, the spin J must be an integer for even-A nuclides and half-integer for odd-A nuclides.

A survey of the spins of stable nuclei in their ground states reveals several important trends. First, even Z-even N nuclides invariably have $J = 0$. This fact along with other evidence suggests that pairs of identical nucleons orient their angular momenta so the sum of the pair is zero. For odd Z-even N or even Z-odd N nuclei (hereafter referred to as "odd-even" nuclides), a variety of experimental evidence suggests that the nuclear angular momentum is usually contributed exclusively by the one unpaired nucleon. That is, at least as far as angular momentum is concerned, we can picture odd-even nuclei as consisting of an even-even spin-zero core plus a single nucleon. For odd Z-odd N nuclei, J is expected to be the vector sum of the angular momenta of the unpaired proton and nucleon, j_p and j_n:

$$J = j_p + j_n$$

Rules of angular momentum addition require in this case that J be one of the following integers:

$$J = |j_p - j_n|, |j_p - j_n| + 1, \cdots, j_p + j_n$$

D. Parity

The wave function $\psi(x, y, z)$ representing a single nucleon is said to have even (or positive) parity if it does not change sign when x is changed to $-x$, y to $-y$, and z to $-z$. If $\psi(x, y, z)$ does change sign, it is said to have odd (or negative) parity:

$$\psi(x, y, z) = \psi(-x, -y, -z) \qquad \text{even parity}$$

$$\psi(x, y, z) = -\psi(-x, -y, -z) \qquad \text{odd parity}$$

It can be shown that there is a relation between the parity of a particle's wave function and the orbital angular momentum quantum number ℓ of the particle:

$$\text{Parity} = (-1)^{\ell}$$

where $+1$ is equivalent to even parity and -1 to odd parity.

For the nucleus, its wave function can be written as the product (or as the sum of products) of the wave functions of the individual nucleons. So the parity of the nucleus is given by the product of the parities of the constituent nucleons. As mentioned in Section II.C, in even Z-even N nuclei in the ground state, nucleons tend to pair their angular momenta to zero. The effect is to cancel the ℓ vectors of each pair so that each pair contributes even parity to the nucleus. Consequently, even Z-even N nuclei have even parity in the ground state. Then we expect the parity of odd-even nuclei to be given by the parity of the single unpaired nucleon. Odd Z-odd N nuclei have a parity which is the product of the parities of the unpaired neutron and proton:

$$\text{Parity} = (-1)^{\ell_p + \ell_n}$$

Customarily, the spin and parity of a state are specified as, for example, $5/2^+$ meaning $J = 5/2$ with even parity.

E. Magnetic Moments

In classical physics, a magnetic dipole moment is produced by a charged object moving in an orbit or spinning about an axis. As expected, then, the magnetic dipole moment of a proton is contributed both by its orbital motion ℓ and by its intrinsic spin s. Neutrons, having no charge, have no magnetic moment by virtue of their orbital motion but, surprisingly, have a magnetic moment due to their intrinsic spin. The dipole moments of the proton and neutron due to intrinsic spin are 2.7925 and -1.9128 nuclear magnetons, respectively.

The total magnetic dipole moment of the nucleus then depends on the orbital angular momenta of the protons and on the intrinsic spins of the protons and neutrons. If it is true that the angular momenta and hence magnetic moments of identical nucleons cancel in pairs, then even Z-even N nuclei should have zero magnetic moments. This expectation is confirmed by experiment.

For odd-even nuclei, the magnetic moment is expected to be due to the angular momentum of the unpaired nucleon. The magnetic moment will depend on whether the ℓ and s vectors of the nucleon are pointing in the same direction (the case of $J = \ell + \frac{1}{2}$) or in opposite directions ($J = \ell - \frac{1}{2}$). For either an unpaired proton or neutron it can be shown that the expression for the magnetic moment μ is (in nuclear magnetons)

$$\mu = \begin{cases} (J - \frac{1}{2}) \, g_{\ell} + \mu_i & J = \ell + \frac{1}{2} \\ \dfrac{J}{J+1} (J + 3/2) \, g_{\ell} - \mu_i & J = \ell - \frac{1}{2} \end{cases} \tag{3}$$

where g_i is 1 for protons and zero for neutrons, and μ_i is the magnetic moment of the proton or neutron due to intrinsic spin. Here, experimental results diverge significantly from what is expected. Data usually fall somewhere between, but rarely on, the predicted values of Equation 3. Apparently, the simple theory on which these predictions are based requires modification.

As for magnetic moments of order higher than the dipole, it can be shown quite generally that nuclei have zero magnetic moments of even order (magnetic quadrupole, 2^4-pole, etc.) because of the definite parity of their wave functions.

F. Electric Moments

It can be shown that the definite parity of nuclear wave functions causes all odd orders of electric moments (dipole, octupole, etc.) to vanish. We wish then to consider the electric quadrupole moment Q which in classical physics is defined for a charge distribution as

$$Q = \int [3z^2 - (x^2 + y^2 + z^2)] \, \rho \, dx \, dy \, dz \tag{4}$$

where ϱ is the charge density at any point (x, y, z). The quantity Q is related to the deviation of the charge distribution from a sphere, as can be seen if we consider a spheroid formed by rotating an ellipse about one of its axes. Suppose the ellipse has semiaxis a along the z direction, which is the axis of rotation, and semiaxis b perpendicular to z. Then if a charge C is uniformly distributed throughout the volume V, the charge density is given by

$$\rho = \frac{C}{V} = \frac{C}{4/3\pi a \, b^2}$$

Performing the integration in Equation 4, we find

$$Q = \frac{2}{5} C (a^2 - b^2)$$

Thus, in the case of a sphere (a = b), Q = 0. When the charge distribution is stretched in the z-direction (a > b), Q is positive. A negative value of Q corresponds to a spheroid flattened along the z axis (a < b).

In the quantum mechanical definition of Q, a fixed direction in space is defined by an electric field, for example, and ϱ is replaced by the proton probability density Ze $|\psi|^2$ where e is the proton charge. Also included in the definition is the requirement that the nucleus be in the state in which the angular momentum J has component J along the fixed direction. The body axis of the nucleus will precess about the fixed direction in space with the result that to an external observer the nucleus will appear to be more spherical than it actually is. In fact, it can be shown that nuclei of spin 0 and ½ have no measurable quadrupole moment even though they may in fact be non-spherical. Thus all even Z-even N nuclei have Q = 0. The quantum mechanical Q also differs in that, customarily, it is given dimensions of distance-squared by including a factor of 1/e in the defining expression.

We wish now to test the hypothesis that odd-even nuclei can be pictured as a spin-zero core plus a single unpaired nucleon. If this were the case, we would expect the quadrupole moment to be due to the odd nucleon alone. Calculations show that the quadrupole moment due to a single proton is

$$Q_p = - \frac{2J - 1}{2(J + 1)} \overline{r^2} \tag{5}$$

where $\overline{r^2}$ is the mean of the square of the distance of the proton from the center of the nucleus. Assuming a value of $\overline{r^2}$ of the same order of magnitude as the square of the nuclear radius, we have from Equation 1:

$$\overline{r^2} \approx A^{2/3} \text{ fm}^2$$

Hence, we expect Q_p to have values of the order of -10 fm^2. Instead, experimental values of Q_p for odd Z-even N nuclei are generally positive and as much as ten times greater in magnitude than expected. Furthermore, even Z-odd N nuclei, which have an unpaired neutron, are expected to have a quadrupole moment Q_n given by

$$Q_n = \frac{Z_2}{A} Q_p \tag{6}$$

that is due to a slight displacement of the center of charge of the nucleus from the center of mass. Instead, odd-N nuclei have values of Q_n very much like the odd-Z nuclides. We conclude that nuclei in general have many protons moving collectively in nonspherical distributions which, if spheroidal, must be elongated along the z-axis of the nucleus.

At certain values of N and Z, the quadrupole moments of odd-even nuclei become suddenly small and negative, as predicted by Equations 5 and 6. These are the magic number-plus-one nuclei which apparently do behave as a spherically symmetric core plus an additional nucleon moving predominately in the xy plane.

III. THE TWO-NUCLEON FORCE

Because of the complexity of nuclear interactions, a detailed knowledge of the nature of nuclear forces is somewhat irrelevant to an introductory discussion of nuclear structure. But the subject of nuclear forces is so central to the whole of nuclear physics that some mention of it is desirable. Also a discussion of this subject will allow certain topics to be introduced which will be useful later.

The simplest nuclear systems are the neutron-proton (the deuteron), the proton-proton (the di-proton), and the neutron-neutron (the di-neutron). Of these, only the deuteron is stable. The experimental study of these systems involves the measurement of the properties of the deuteron ground state or the scattering of a beam of protons or neutrons from a target of protons with the measurement of the intensity of the scattered beam as a function of angle. These observations yield the most accurate information we have about nuclear forces. Whether knowledge of the two-nucleon force can be carried over to systems of greater mass number is unknown, however, since the presence of other nucleons may modify the effect of the two-nucleon interaction.

There is a fundamental theory of the nuclear force, the *meson field theory,* which at least in its essential assumptions is thought to be correct. This theory postulates that nuclear forces arise between nucleons because of the exchange of particles called mesons. This picture is given credence because electromagnetic forces, which are well understood, are known to be associated with the exchange of photons between charged particles. However, whereas the theory for electromagnetic interactions is mathematically tractable, the meson theory suffers from various technical problems of mathematics and physics.

In the absence of a satisfactory fundamental theory, one attempts to find a suitable force law by testing various alternatives against the pertinent experimental data. Fortunately, the number of possibilities is greatly limited by the conservation principles of energy, linear and angular momentum, parity, and charge. It can be shown that the most general nonrelativistic potential V (relativistic effects are important only in high energy scattering experiments) satisfying the conservation principles has the form:

$$V = V_{\text{ordinary}} + V_{\text{exchange}} + V_{\text{tensor}} + V_{\text{velocity-dependent}} \tag{7}$$

Each of the components of Equation 7 will now be discussed.

A. $V_{ordinary}$

A central force between two particles is one which acts along the line joining them. Gravitational and electrostatic forces have this property. $V_{ordinary}$ represents the potential energy associated with a central force that depends only on the distance r between the nucleons. Such a force is called a *Wigner force*. We expect the Wigner force to be attractive, at least in part, since in the absence of an attractive nuclear force the protons would fly apart because of Coulomb repulsion. Experiment shows in addition that the Wigner force is of short range. A nucleon projected at a nucleus feels almost no nuclear force when it is farther than 2 or 3 fm from the nuclear surface.

An examination of the deuteron provides some information about $V_{ordinary}$. The observed properties of the deuteron ground state are limited to the following:

- Binding energy = 2.22 MeV
- Spin and parity = 1^+
- Quadrupole moment = $2.73 \times 10^{-27} cm^2$
- Magnetic moment = 0.8574 nuclear magnetons

If we take $V_{ordinary}$ to be zero outside the range of nuclear forces, then inside this range it must become negative corresponding to an attractive force. However, the exact shape of $V_{ordinary}$ cannot be derived from the limited number of properties of the deuteron. If for purposes of illustration we use a square well potential of depth V_o and width R, then V_o is about 30 MeV if R is taken to be about 2 fm. This rough idea of the nuclear potential energy illustrates the strength of the nuclear force since the electrostatic potential energy of two protons located 2 fm apart is 0.7 MeV and the gravitational potential energy is 6×10^{-37} MeV under these circumstances.

The scattering of high energy protons from protons shows that the Wigner force also includes a repulsive component which becomes important when two nucleons approach closer than about $\frac{1}{2}$ fm. The height of the repulsive part of $V_{ordinary}$ is very great, perhaps infinitely high, as shown in Figure 1. It is as though each nucleon has at its center a *repulsive core* which in effect makes it impenetrable. A repulsive core helps to explain a basic property of nuclei, namely, that their density is approximately constant regardless of mass number. If nuclear forces were only attractive, one would expect the nucleus to collapse to a radius such that every nucleon would exist within the range of force of every other. Although the kinetic energy of the nucleons would increase under these circumstances, this increase would be insufficient to prevent collapse. Instead, as mentioned in Section II.A, measurements of nuclear density show that the nuclear volume is proportional to the number of nucleons as though each nucleon were an impenetrable sphere.

Another well-known property, the approximate proportionality of nuclear binding energy to the number of nucleons A, is also related to the repulsive core. If every nucleon were bound to every other, as in a collapsed nucleus, the total binding energy would be proportional to the number of pairs of nucleons, which is approximately A^2. Repulsive forces require that nucleons remain separated, resulting in a nucleus of large dimensions compared to the range of internucleon forces. Hence, a nucleon interacts with only a limited number of other nucleons, a phenomenon known as the saturation property of nuclear forces. As described below, exchange forces also act to restrict the binding among nucleons.

B. $V_{Exchange}$

It is possible for forces to exist between two particles which exchange the coordinates of the particles. For example, in the H_2 molecule, electrons are exchanged between the

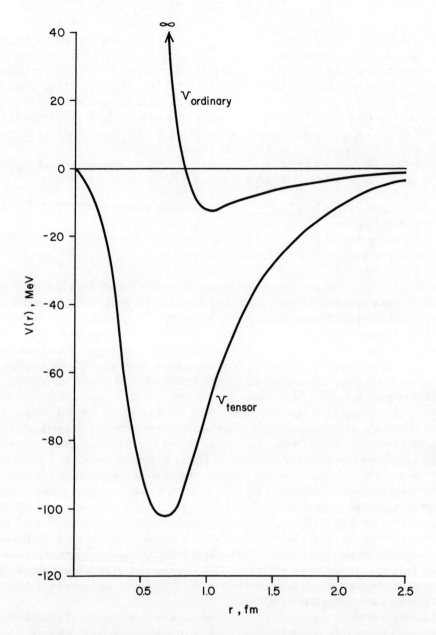

FIGURE 1.   Ordinary and tensor potential energy functions for the deuteron, derived from a meson theory of nuclear forces. (After Gartenhaus, S., *Phys. Rev.*, 100, 903, 1955. With permission.)

two atoms. Like electrons, nucleons have position coordinates and intrinsic spin coordinates. Three distinct types of exchange are possible:

1.   Exchange of position coordinates alone (Majorana exchange). For a two-particle system this kind of exchange is equivalent to the parity inversion operation. If the two particles are in a state of even orbital angular momentum L, the wave function of the system will be unchanged under Majorana exchange. If L is odd, the wave function changes sign.

2.   Exchange of spin coordinates alone (Bartlett exchange). We recall that when two particles have their spins parallel (triplet state), the system wave function is unaltered under a spin exchange. For spins aligned antiparallel (singlet state), the wave function changes sign.
3.   Exchange of both position and spin coordinates (Heisenberg exchange). Heisenberg exchange is equivalent to a Majorana plus a Bartlett exchange. Thus, a Heisenberg exchange does not alter the wave function of L-even, triplet state systems or L-odd, singlet state two-body systems. The sign of the wave function changes for L-even singlet state or L-odd triplet state systems.

In general, a potential corresponding to each of these exchange types is possible:

$$V_{exchange} = V_{Majorana} + V_{Bartlett} + V_{Heisenberg}$$

where the Vs are functions of the nucleon separation distance r and have negative or positive signs depending on whether the particular type of exchange does or does not change the sign of the system wave function, respectively.

Which if any of these exchange potentials is important will now be considered. The ground state of the deuteron is found to be a triplet state and no stable singlet state exists. This fact suggests that nuclear forces are spin dependent since there is no other reason why a bound singlet state should not exist. Apparently, the nuclear force for the triplet state is more strongly attractive than that for the singlet state. Indeed, experiments in which low energy neutrons are scattered from protons show that the well depth of the triplet state potential energy is about three times as great as that for the singlet state. This spin-dependent effect can be accounted for by either Bartlett or Heisenberg exchange forces, or both. The range of these forces is comparable to that of the Wigner force.

The existence of a Majorana exchange force is indicated by the results of scattering high energy neutrons from protons. Because of their high energy relative to the depth of the nuclear potential well, one would expect the neutrons to be little deflected. Instead, strong backward scattering is observed in a coordinate system in which the center of mass of the neutron and proton remains stationary. This fact suggests that incident neutrons have exchanged positions with protons in the target so that protons go forward and neutrons backward. Like the repulsive core, Majorana exchange forces contribute to the saturation property of nuclear forces because whenever the exchange of a pair of nucleons in a nucleus changes the sign of the wave function, the force between the nucleons will be repulsive thus reducing nuclear binding.

Majorana exchange also plays an important role in keeping the number of protons approximately equal to the number of neutrons in stable nuclides. To understand this fact in a simplified way, we assume that nucleons move independently of each other in a simple attractive potential. A solution of Schrödinger's equation for the potential yields a series of energy levels. Beginning with the lowest in energy, each level is filled with the maximum number of nucleons. Since the Pauli exclusion principle applies separately to neutrons and to protons, there can be two neutrons and two protons in each state. Each pair has its spins oriented antiparallel corresponding to the antisymmetric spin state and the symmetric space state. That is, the exchange of the space coordinates of identical nucleons in the same state will not change the sign of the wave function and so the Majorana exchange force will be attractive. Consider, however, the exchange of two identical nucleons having the same spin orientation but residing in different levels. An exchange of these identical particles will alter the sign of the wave function, corresponding to a repulsive Majorana force. Exchange of nonidentical nucleons in different levels can be shown to not affect the binding energy. Consequently, in order to maximize the amount of attractive Majorana exchange force, the number of nucleons in equal levels must be maximized. To put it another way, the total number of occupied levels must be minimized. This occurs when the number of protons is approximately equal to the number of neutrons.

C. V_{tensor}

The nuclear force contains a noncentral component which is known as the tensor force, although V_{tensor} is actually a scalar quantity formed from two tensors. That a noncentral force exists is known from the existence of an electric quadrupole moment for the deuteron. It can be shown that when a central, attractive force exists between two particles that the lowest possible energy state for the system corresponds to a total orbital angular momentum of zero. A characteristic of L = 0 states (S states) is that their probability density is spherically symmetric. We now inquire whether the ground state of the deuteron is an S state. One way of experimentally answering this question is to measure the quadrupole moment of the deuteron. The existence of a nonzero quadrupole moment indicates that the deuteron is not entirely in an S state. Although the existence of a nonzero quadrupole moment does not necessarily imply the presence of noncentral forces in nuclei having more than two nucleons, it unquestionably does for the deuteron.

If we ignore velocity-dependent effects, the only preferred directions in space for the deuteron are the directions of spin s_1 and s_2 of the proton and neutron and the direction of the line joining the nucleons r. Thus a noncentral force can only depend on the angle between the two spin directions or on the angle between the spin directions and r. The most general form for a noncentral potential which satisfies the conservation principles is

$$V_{tensor} = V_T(r) \ (3 \, \sigma_1 \cdot \hat{r} \, \sigma_2 \cdot \hat{r} - \sigma_1 \cdot \sigma_2)$$

where $V_T(r)$ is an ordinary potential function of the distance between nucleons, r is the unit vector in the direction of r, and $\sigma = 2s$. The effect of V_{tensor} is to add to the S state of the deuteron small contributions of states of higher L. These states must all have the same parity, if parity is to be definite for nuclei, so that admixtures of L = 2, 4, 6, $\cdots$ states are possible in the deuteron. Calculations show that the deuteron ground state is 96% S state and 4% D state (L = 2), both triplet. The fact that there is little admixture to the S state, however, does not imply that the tensor force is relatively weak. On the contrary, it is believed that the tensor force in the triplet state is of considerably greater strength and of slightly greater range than the central force, as shown in Figure 1. However, in singlet states V_{tensor} vanishes.

D. $V_{velocity\text{-}dependent}$

In principle, velocity-dependent forces are possible which depend on the relative linear momentum p of two nucleons or on their orbital angular momentum L = r × p. Several lines of evidence suggest that at least one velocity dependent force exists for nucleons, the spin-orbit force, in which the orbital angular momentum L interacts with the intrinsic spin angular momentum S of the nucleons. The form of the spin-orbit potential is

$$V_{velocity\text{-}dependent} = V_{S\text{-}O} \ \mathbf{L} \cdot \mathbf{S}$$

where $V_{s\text{-}o}$ is a function only of the separation distance between nucleons. The range of $V_{s\text{-}o}$ is thought to be very short.

These briefly are the known properties of the nuclear force. As an approximation one may assume that the Wigner, Majorana, Bartlett, and Heisenberg potentials all have the same r-dependence and ask what strengths of these potentials give the best fit to nuclear structure and scattering data. It is found that equally good fits can be made with different sets of strengths so that it is seemingly impossible to determine the individual contributions uniquely. More sophisticated potentials based on meson theory provide more satisfactory fits of data but are too difficult to consider here.

It has been tacitly assumed that the nuclear force does not depend on whether the nucleons are protons or neutrons. We now wish to consider this topic briefly. The meson-exchange theory of nuclear forces predicts that the n-n force is identical to the p-p force but that the n-p force may differ slightly from these. In the event that n-n and p-p forces are identical without regard to the n-p force, we say that nuclear forces are *charge symmetric*. If n-p, n-n, and p-p forces are all identical, we say that nuclear forces are *charge independent*.

At first glance it would seem that the existence of a stable deuteron system but not of a stable dineutron or diproton system argues against the charge independence of nuclear forces. However, the Pauli exclusion principle forbids two protons or neutrons from forming a triplet S state, which is the ground state of the deuteron. As we have seen, nuclear forces are spin-dependent and so one would have to compare singlet S states of the n-p, n-n, and p-p systems, all of which turn out to be unbound. A better test is afforded by the study of *mirror nuclei* which are two nuclei for which the N, Z-values of one are equal to the Z, N-values of the other. The simplest example is $^3_1$H and $^3_2$He where ^{3}H has one n-n force and two n-p forces while ^{3}He has one p-p force and two n-p forces. If charge symmetry holds (n-n and p-p forces identical), then the binding energy of the two nuclei should be equal if we correct for the contribution due to Coulomb forces in ^{3}He. Calculations of this type show that nuclear forces are charge symmetric, at least to an excellent approximation. Additional evidence from low and high energy nucleon-nucleon scattering indicates that within experimental uncertainty nuclear forces are also charge independent.

Because the nuclear force does not depend on whether nucleons are protons or neutrons, we may consider these particles as being two different states of the same particle, the nucleon. We distinguish formally between the two states by a variable called *isospin* which has mathematical properties analogous to intrinsic spin. The isospin formalism is useful for expressing those nuclear symmetries which are due to the charge independence of nuclear forces.

IV. NUCLEAR STATES

A. Doubly Magic Nuclides and Their Immediate Neighbors

We have seen that when Z or N is equal to one of the magic numbers 2, 8, 20, 28, 50, 82, or 126 that certain nuclear properties take on unusual values. Magic nuclei and in particular *doubly magic nuclei,* which have both Z and N equal to a magic number, are especially tightly bound and have spherical shapes. These physical properties allow a particularly simple theoretical description to be made of nuclei which are doubly magic or only one or two nucleons removed from this status. The fact that magic nuclei are spherical implies that insofar as nucleons move independently of one another (a point to be considered below), the potential in which they move is spherically symmetric. The stability of doubly magic nuclei suggests that the addition of one more nucleon may not significantly disturb the inert "core" so that the properties of the nucleus are due to the single nucleon. Similarly, in a doubly magic nucleus from which one nucleon has been removed, the nuclear properties will be determined by the "hole". This is because it can be shown on theoretical grounds that the existence of a hole in a closed shell configuration is equivalent to having a single particle in the shell with the same angular momentum as the hole. Indeed, as we have seen, the ground state spin and parity, the electric quadrupole moment, and to a lesser extent the magnetic dipole moment are determined by the properties of the "valence" nucleon or hole.

The very fact that magic numbers exist suggests that each nucleon moves more or less independently of the others. If this were not the case, one would expect the energies and angular momenta of nucleons to be constantly changing because of collisions so that their motion would be chaotic and unrelated to N or Z. On the other hand, it is difficult at first glance to see why nucleons should move independently when the forces between them are short range and very strong. It is the Pauli exclusion principle which resolves this difficulty. Only one nucleon at a time can be in a given state and under normal circumstances all the lowest states are filled. In a collision two particles must transfer energy, but because of the dearth of available alternative states to occupy they are effectively prevented from interacting.

In order to express in mathematical form the properties of nuclei near the magic numbers, we make the fundamental assumption that the force on each nucleon depends only on the distance r of the nucleon from the center of the nucleus. Contained within this assumption is the idea that internucleon interactions average out entirely so that each nucleon experiences the presence of others only indirectly through a force field that does not change with time. The r-dependence of the force and associated potential energy is a result of the fact that nuclei near magic numbers are approximately spherical. Because of the conservation of orbital angular momentum of a particle moving in a central force field, nucleons are expected to have definite orbital angular momenta l and total angular momenta j.

These assumptions form the basis of the *single particle model*. Because of its similarity to the description of atomic electrons and their shell structure, this theory or more sophisticated versions of it may be referred to as the *shell model* of the nucleus.

The shape of the potential energy V(r) associated with the force felt by each nucleon will now be considered. Since internucleon forces are attractive and short range, we expect that V(r) will be lowest in the interior of the nucleus, rising to a maximum at the nuclear surface. In fact, we expect the shape of V(r) to approximately resemble the nuclear density distribution (Figure 2) but with opposite sign.

For illustrative purposes we use a parabolic shape for V(r):

$$V(r) = \tfrac{1}{2}kr^2 \qquad\qquad (8)$$

where k is a constant. This choice is referred to as the *harmonic oscillator potential* since a mass on a spring feels a force proportional to the distance from its equilibrium position and has a potential energy proportional to the square of this distance. The constant k in Equation 8 is the so-called spring constant and is related to the mass m and frequency of vibration ν in the mass-on-a-spring system by

$$\nu = \frac{1}{2\pi} \sqrt{\frac{k}{m}}$$

In addition to its mathematical simplicity, a potential of this type has the advantage of being applicable to nonspherical situations, a characteristic which will be useful in the study of deformed nuclei. This extension can be made by writing Equation 8 in terms of rectangular coordinates and by allowing a different spring constant k or frequency of oscillation ν in each coordinate direction:

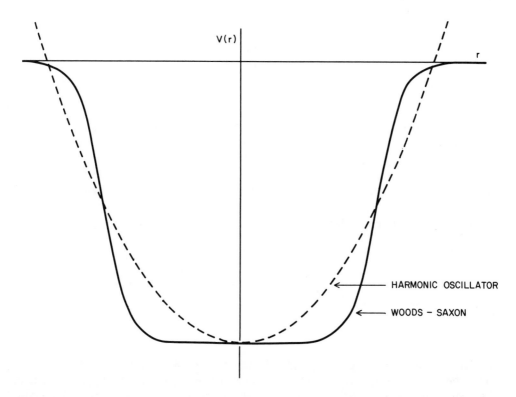

FIGURE 2. Two possible potential energy functions for a nucleon in the nucleus: (A) the Woods-Saxon potential ($V = -[1 + \exp(r - a)/b]^{-1}$) which resembles the nuclear density distribution but with opposite sign; and (B) the simple harmonic oscillator potential ($V = \frac{1}{2}kr^2$) which is more mathematically tractable.

$$V(x, y, z) = \frac{1}{2}(k_x x^2 + k_y y^2 + k_z z^2) \tag{9}$$

Since the nucleons constitute a quantum mechanical system, the permissible energies which they may have are found from the substitution of V into the time-independent Schrödinger equation. For the potential of Equation 9, the energy eigenvalues are given by

$$E = (n_x + \tfrac{1}{2})\,h\nu_x + (n_y + \tfrac{1}{2})\,h\nu_y + (n_z + \tfrac{1}{2})\,h\nu_z$$

where $\nu_x = \dfrac{1}{2\pi}\sqrt{k_x/m}$, etc. The numbers n_x, n_y, n_z can take on the values 0, 1, 2, 3, $\cdots$, and can be thought of as the number of energy quanta required to excite a nucleon to one of the permissible energy states.

For spherical nuclei,

$$\nu_x = \nu_y = \nu_z = \nu$$

and

$$N = n_x + n_y + n_z \tag{10}$$

so that

$$E = (N + 3/2)\,h\nu$$

where $N = 0, 1, 2, 3, \cdots$. Nucleons in a spherical nucleus then may have energies equal to $3/2\ h\nu$, $5/2\ h\nu$, $7/2\ h\nu$, $\cdots$. Nucleons have in addition to an energy E an orbital angular momentum l with magnitude

$$|l| = \sqrt{l(l+1)}\ \hbar$$

where the orbital angular momentum quantum number l may take on the values

$$l = 0, 1, 2, 3, 4, \cdots$$

As in atomic spectroscopy, these values of l may equivalently be specified by the letters s, p, d, f, g, $\cdots$.

It can be shown that a relation exists between N and l such that

$$l \leqslant N$$

and even l corresponds to even N, odd l to odd N. The usual association between l and the parity of the state exists

$$\text{Parity} = (-1)^{l}$$

so that even-N states have even parity and odd-N states have odd parity. Customarily, one labels each energy level by either its value of N or by the l value that the nucleon has. Since nucleons in different energy levels may have the same l value, one distinguishes between the s states, for example, by labeling them 1s, 2s, 3s, $\cdots$, where 1s means the first s state, etc. Thus the $N = 0$ energy level contains 1s nucleons, the $N = 1$ level contains 1p nucleons, the $N = 2$ level contains 2s and 1d nucleons, etc., as shown in Figure 3.

To determine how many nucleons may exist in each energy level, we recall the Pauli exclusion principle which states that no two identical fermions (protons or neutrons in this case) can have exactly the same quantum numbers. For the harmonic oscillator potential, the quantum numbers are N, l, and in addition m_l, the quantum number giving the various possible components of l along a fixed direction in space

$$m_l = -l, -l+1, \cdots, l$$

and m_s, the quantum number giving the two components of the intrinsic spin s along the fixed direction

$$m_s = -\tfrac{1}{2}, \tfrac{1}{2}$$

The total number of protons or neutrons in each energy level turns out to be $(N + 1)(N + 2)$. For the harmonic oscillator potential the energy levels are evenly spaced. We would expect that if the single particle model applies to all nuclei, then a significant energy gap would appear after neutron or proton numbers of 2, 8, 20, 40, 70, 112, and 168, as shown in Figure 3. Nuclei with N or Z equal to one of these numbers would be comparatively tightly bound since additional nucleons must go into higher energy states. That is, the single particle model assumptions along with a harmonic oscillator potential predict these as the magic numbers. Unfortunately, only the first three correspond to the shell closures indicated by experiment, where the term *shell* means those nucleon states which are grouped together in energy between adjacent magic numbers.

17

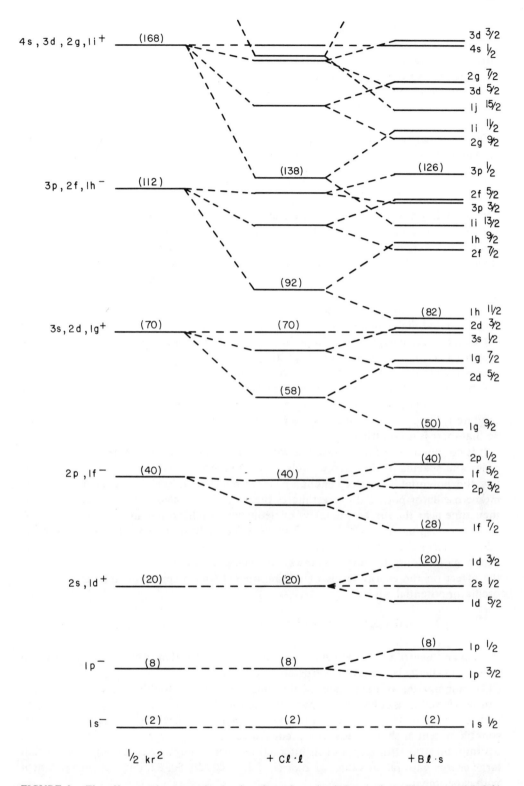

FIGURE 3. The effect on the energy levels of nucleons in a simple harmonic oscillator potential (left) from the addition of a $C\boldsymbol{l}\cdot\boldsymbol{l}$ interaction (middle) and a $B\boldsymbol{l}\cdot\mathbf{s}$ spin-orbit interaction (right). Numbers in parentheses are the magic numbers predicted by each potential. (After Nilsson, S. G., *Dan Mat. Fys. Medd.*, 29 (16), 1955. With permission.)

Although it might be thought that a better choice of potential energy function would reproduce the true magic numbers, it happens that the shell closures are rather insensitive to the shape of the potential well. However, if a strong spin-orbit interaction is hypothesized so that a term proportional to l·s is added to the potential of Equation 8, the magic numbers can be readily attained. This key to the "periodic table of the nuclides" was discovered by Mayer[1] and by Haxel et al.[2] In the spin-orbit interaction the intrinsic spin s of a particle is coupled with its orbital angular momentum l to form a total angular momentum j. Because the intrinsic spin quantum number s has only the value one half, j is restricted to the values

$$j = \ell - \tfrac{1}{2},\ \ell + \tfrac{1}{2}$$

corresponding to the cases in which *l* and s are either pointing in the opposite or in the same directions. To obtain the magic numbers one finds that the energy of a nucleon must increase when j = *l* − ½ and decrease when j = *l* + ½. The spin-orbit interaction for electrons, which is due to electromagnetic rather than to nuclear forces, has the opposite effect and is much weaker. For both nucleons and electrons it can be shown that the difference in energy (the "splitting") between the j = *l* + ½ and j = *l* − ½ cases is proportional to (2*l* + 1). Thus, the splitting is largest for states of large *l* so that high angular momentum states are radically shifted in position to change the order of the energy levels. In addition, the number of states of different energy is increased because of the spin-orbit splitting. The proper quantum numbers for labeling each state are now N, *l*, j, and m_j. Conventionally, the j quantum number appears as a subscript in the level designation, as in $1d_{5/2}$. Since m_j can take on any value from −j to +j in integer steps, the number of neutrons or protons that each state can accommodate is (2j + 1). Special cases are *l* = 0 states for which j can only equal one half so that there is no splitting.

It is not possible to exactly produce with the harmonic oscillator potential the correct order and spacing of the energy levels as determined experimentally. The difficulty is that this potential is not deep enough near the nuclear surface. Consequently, the harmonic oscillator potential overestimates the energy of nucleons which spend much of their time near the surface, namely, nucleons with high orbital angular momentum *l*. Better correspondence with reality can be effected in a somewhat artificial way by adding a term C *l·l* to the potential energy function of Equation 8. With C taken to be a negative constant, this term lowers the energy of high-*l* nucleons. Figure 3 shows the effect on the energy levels of adding an *l·l* and a spin-orbit interaction to the oscillator potential, giving

$$V(r) = \tfrac{1}{2} kr^2 + B\ell \cdot s + C\ell \cdot \ell \tag{11}$$

Proton and neutron levels are essentially the same except that proton levels are shifted upward in energy because of electrostatic repulsion.

To summarize, we have assumed that nuclei which are doubly magic, or nearly so, can be viewed as a collection of particles moving independently in a spherically symmetric potential. The solution of the time-independent Schrödinger equation for a reasonable potential shape leads to a predicted set of energy levels. With the additional assumption of a strong spin-orbit interaction, these energy levels are ordered so that large energy gaps fall at values of Z or N of 2, 8, 20, 28, 50, 82, and 126, in agreement with experiment.

Surprisingly, with one additional assumption, the single particle model is able to predict the ground state spins and parities of nearly all nuclides. The additional assumption is that neutrons and protons separately cancel their angular momenta in pairs. This pairing phenomenon has no explanation in the single particle model and in fact is a violation of the hypothesis that no internucleon forces exist in addition to the average central force. Nevertheless, the existence of *pairing forces* is indicated by the fact that all even Z-even N nuclei have zero spin ground states. This fact is unexpected since the total angular momentum of an even number of nucleons can take on a variety of nonzero values in general. In a semiclassical argument let us assume that pairing forces are short range and attractive so that two nucleons are most tightly bound together when they are closest to one another. This arrangement is enhanced when the nucleons are orbiting in the same plane but in opposite directions so that they pass each other frequently. Thus pairing forces cause the l vectors of a pair of nucleons to point in opposite directions and to cancel the total angular momentum of the pair. An additional effect of pairing forces is to require a large amount of energy to raise one member of a nucleon pair to a higher level since energy is needed to split the pair. It happens that this pairing energy increases with increasing l of the nucleons so that pairs of particles tend to occupy high-l states.

Ground state spins and parities are expected to obey the following rules:

1. Even Z-even N nuclides have zero spin and positive parity.
2. Odd Z-even N or even Z-odd N nuclides have the spin and parity of the unpaired nucleon. These values are found in the following way. For an odd-N nuclide, for example, using Figure 3 one fills with neutrons the lowest states first, each state having a maximum of $(2j + 1)$ particles, until the state containing the last neutron is reached. The j and l of this state are noted. Then j is the nuclear spin and the parity is given by $(-1)^l$. Exceptions to this rule may occur when the state of the unpaired nucleon has many neighbors close in energy or when the nucleus is far from doubly magic.
3. Odd Z-odd N nuclides present a more complicated case since then there are two unpaired nucleons whose angular momenta j_1 and j_2 may combine in several different ways. According to the rules of angular momentum addition, if

$$J = j_1 + j_2$$

then the quantum number J may take on the values

$$J = |j_1 - j_2|, |j_1 - j_2| + 1, \cdots, j_1 + j_2$$

The parity will be the product of the neutron and proton parities:

$$\text{Parity} = (-1)^{l_1 + l_2}$$

For most odd-odd nuclei, *Nordheim's rules* apply. If we define Nordheim's number N as

$$N = j_1 + j_2 - (l_1 + l_2)$$

then the spin J of an odd-odd nuclide is predicted to be:

$$\text{when } N = 0 \quad J = |j_1 - j_2|$$

$$\text{when } N = \pm 1 \quad J \text{ is either } |j_1 - j_2| \text{ or } |j_1 + j_2|$$

Nordheim's rules can be theoretically justified on the basis of spin-dependent forces.

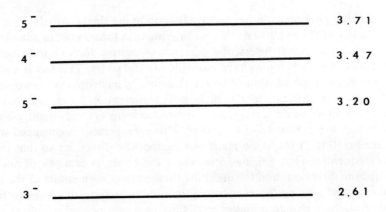

5⁻ ————————————————————— 3.71

4⁻ ————————————————————— 3.47

5⁻ ————————————————————— 3.20

3⁻ ————————————————————— 2.61

0⁺ ————————————————————— 0 MeV

$^{208}_{82}$Pb

FIGURE 4.   Low-lying energy levels of the doubly-magic $^{208}_{82}$Pb nucleus.

Whereas the single particle model is quite successful in predicting the ground state spins and parities of most nuclides, its predictions of the properties of excited states are accurate only for nearly doubly magic nuclides. Several examples follow.

Example 1 — $^{208}_{82}$Pb. With Z = 82 and N = 126, ^{208}Pb is an excellent example of the "inert core", doubly magic nuclide. In Figure 4 are shown its four lowest excited states. The energy of the first excited state, a 3⁻ level at 2.61 MeV, is higher than that of any other medium or high-A nuclide, a fact which demonstrates the inherent stability of this nucleus.

One expects the low-lying excited states of ^{208}Pb to result from the promotion of

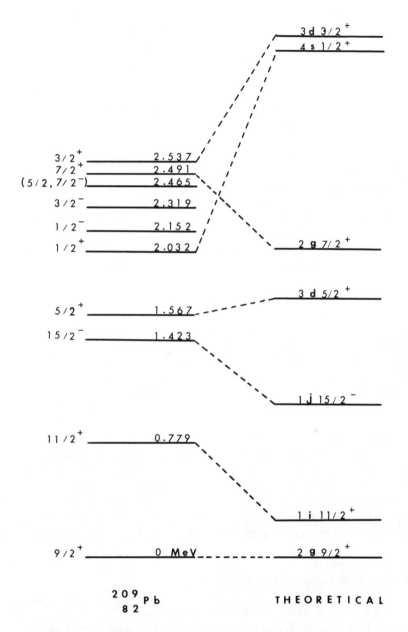

FIGURE 5. Experimentally observed energy levels of $^{209}_{82}$Pb (left) compared
with the levels predicted by the single particle model of the nucleus (right).

one of its protons or neutrons to a higher level. When the nucleon is promoted, a hole
is left behind in the closed shell which behaves as a single particle. The nucleon and
hole each have angular momenta that combine by the usual rules to give the total
angular momentum of the nucleus. To see how this proceeds we refer to Figure 3 where
the levels above $Z = 82$ and $N = 126$ can be seen. The elevation of a $p_{1/2}$ neutron to
the $g_{9/2}$ level, for example, can result in excited states having either of two spins: $(9/2
- 1/2) = 4$ or $(9/2 + 1/2) = 5$. In either case the parity is determined by the l-values
of the particle and hole: parity $= (-1)^{2+5} = -1$. The situation is complicated by the
fact that several other nucleon-hole configurations can also give rise to 4$^-$ and 5$^-$ states.
In general, each observed nuclear state contains contributions from more than one

single particle excitation. Gamma ray transition rates and other evidence show that the 3⁻ first excited state is not due to single particle excitations but to a vibrational motion of the nucleus which is discussed in Section IV.B. Other excited states of ^{208}Pb can, however, be identified with the excitations of single nucleons. According to a theoretical analysis[3] the levels at 3.20 and 3.47 MeV are due largely to a $g_{9/2}$ neutron - $p_{1/2}$ hole configuration, while the state at 3.71 MeV has contributions from $g_{9/2}$ neutron - $f_{5/2}$ hole and $h_{9/2}$ proton - $s_{1/2}$ hole configurations.

Example 2 — $^{209}_{82}$Pb. The addition of a neutron to the doubly magic ^{208}Pb core gives ^{209}Pb. We expect that the unpaired neutron can be promoted to higher levels without disturbing the core so that the spin, parity, and energy of the nucleus are contributed solely by the single neutron. Experiment shows this to be largely the case. In Figure 5 the observed energy level scheme is compared with the energy levels above N = 126 predicted by the harmonic oscillator potential with spin-orbit and $\mathit{l} \cdot \mathit{l}$ interactions. Although the level spacings differ, there is general agreement between the two. It is also possible for the single neutron to interact with vibrational motion of the core as described in Section IV.B. This interaction gives rise to the states at 2.152, 2.319, and 2.465 MeV.[4]

Example 3 — $^{210}_{82}$Pb. With two neutrons added to the ^{208}Pb core, low-lying excited states of ^{210}Pb are expected to result from the interaction of the two neutrons in their $g_{9/2}$ state. States can also be formed from the promotion of one extracore neutron to an excited level. A third source of excited states is the interaction of the two neutrons when they have been promoted to the next higher level, the $i_{11/2}$. In each of these cases, the spins of the states so formed are determined by the various ways in which the angular momenta of the neutrons or of the neutron and hole can couple. The interaction of two $g_{9/2}$ neutrons is expected to produce states of angular momenta ranging from $(9/2 - 9/2) = 0$ to $(9/2 + 9/2) = 9$ with even parity. Because these are identical fermions with the same N, l, and j values, it can be shown that only even values of angular momentum are permitted so that the possible states are 0, 2, 4, $\cdots$, 8⁺. Similarly, the interaction of the neutron pair promoted to the $i_{11/2}$ state produces the excited levels: 0, 2, 4, $\cdots$, 10⁺. States formed by the promotion of a single $g_{9/2}$ neutron to the $i_{11/2}$ state have spins ranging from $(11/2 - 9/2) = 1$ to $(11/2 + 9/2) = 10$ in integral steps with parity given by $(-1)^{4 + 6} = +1$. Similarly, the promotion of a $g_{9/2}$ neutron to the $j_{15/2}$ state can produce the excited states: 3, 4, 5, $\cdots$, 12 of parity $(-1)^{4 + 7} = -1$.

In Figure 6 is shown the energy level scheme of ^{210}Pb and the neutron configurations responsible for these states as predicted by theory.[5]

B. Nuclides Somewhat Removed from Doubly Magic

Let us form a semiclassical picture of the doubly magic nucleus to which have been added (or from which have been subtracted) several nucleons. The closed-shell portion of the nucleus is a spherical core while the nucleons beyond the core (the "extracore" nucleons) tend to have a nonspherical distribution. The effect of the extracore nucleons on the core is to distort its shape, while the core which prefers sphericity attempts to keep the extracore nucleons in as spherical a distribution as possible. Pairing forces between extracore nucleons also act to keep their distribution spherical by causing their angular momenta to pair to zero.

If the number of extracore nucleons is not too large, the nucleus will remain spherical on the average but will vibrate about a spherical shape. Such a phenomenon requires the cooperative motion of many nucleons, and theories which describe these kinds of behavior are called *collective models* of the nucleus. Measurements of various nuclear parameters indicate that collective motions do not involve the entire set of nucleons but occur mainly at the surface. These surface fluctuations occur much more

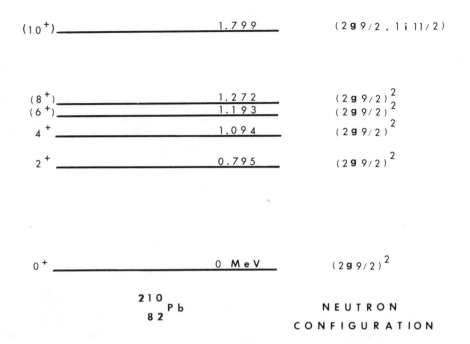

(10^+) _____ 1.799 _____ $(2\,g\,9/2\,,\,1\,i\,11/2\,)$

(8^+) _____ 1.272 _____ $(2\,g\,9/2\,)^2$
(6^+) _____ 1.193 $(2\,g\,9/2\,)^2$
4^+ _____ 1.094 _____ $(2\,g\,9/2\,)^2$

2^+ _____ 0.795 _____ $(2\,g\,9/2\,)^2$

0^+ _____ 0 MeV _____ $(2\,g\,9/2\,)^2$

$^{210}_{82}$Pb

NEUTRON

CONFIGURATION

FIGURE 6. Energy levels of the $^{210}_{82}$Pb nucleus which contains two neutrons in addition to a doubly magic core. The configuration of the two neutrons largely responsible for each level is shown at the right.

slowly than the motion of individual nucleons so that it is possible to treat the dynamics of the two independently to a large extent. A second kind of collective motion, rotation of the nucleus, does not become important until the number of extracore nucleons is large, at which point the nuclear shape becomes permanently distorted. Most nuclides fall into the class which is considered in this section, namely, nuclides intermediate between the permanently spherical and the permanently deformed. Unfortunately, theories applicable to this class are probably the least successful in predicting the properties of excited states. One theory, the vibrational model of Bohr and Mottelson,[6] is presented in detail. Another, more sophisticated theory is the quasiparticle formulation of which an introductory discussion appears in Cohen's *Concepts of Nuclear Physics*.[7]

The problem of mathematically describing small deviations from a spherical shape was treated by Rayleigh in 1877. What is done is to parameterize deviations with the use of spherical harmonics $Y_{\lambda\mu}(\theta, \phi)$ which are a set of mathematical functions whose values depend only on the angles θ and ϕ in spherical coordinates. Members of the set $Y_{\lambda\mu}$ are designated by the integers $\lambda = 1, 2, 3, \cdots$, and by μ which ranges in value from $-\lambda$ to $+\lambda$ in integer steps. The nuclear radius $R(\theta, \phi)$ at any angle θ and ϕ is written in terms of the equilibrium radius R_o as

$$R(\theta, \phi) = R_o [1 + \sum_{\lambda\mu} \alpha_{\lambda\mu} Y_{\lambda\mu}(\theta, \phi)]$$

Hence, $\sum \alpha_{\lambda\mu} Y_{\lambda\mu}(\theta, \phi)$ represents the deviation from equilibrium as a function of angle. For small deviations, the constants $\alpha_{\lambda\mu}$ are small and only a limited number of terms in the summation may be needed.

Vibrational motion has associated with it both kinetic energy T due to the motion of the nucleons and potential energy V associated with the restoring forces of the nuclear surface tension. It can be shown that the energy of vibration E_{vib} can be written in terms of the $\alpha_{\lambda\mu}$ as

$$E_{vib} = T + V$$

$$= \sum_{\lambda\mu} [\tfrac{1}{2} B_\lambda (d\, \alpha_{\lambda\mu}/dt)^2 + \tfrac{1}{2} C_\lambda \alpha_{\lambda\mu}^2] \qquad (12)$$

where B_λ and C_λ are constants that depend on the particular nuclide. Mathematically, Equation 12 is analogous to the energy associated with an assembly of independent simple harmonic oscillators. Each oscillator is capable of being excited to a particular frequency ν_λ, or an integer multiple of ν_λ, that is given by

$$\nu_\lambda = \frac{1}{2\pi} \sqrt{C_\lambda/B_\lambda}$$

The quantum of energy required to excite the fundamental frequency of each type of vibration is called a *phonon* of order λ. Phonons have associated with them not only energy but also an angular momentum λ, with component along the z-axis of μ and parity of $(-1)^\lambda$. Corresponding to each order of vibration λ is a specific class of nuclear shape oscillations. For example, vibrations of $\lambda = 1$ (so-called dipole vibrations) correspond to a simple translation of the center of mass of the nucleus. This type of motion requires an external force on the nucleus and is not of interest here. For vibrations of $\lambda = 2$ (quadrupole vibrations) the nuclear shape remains symmetric about one particular axis, the "principal" axis. Vibrations of $\lambda = 3$ (octupole vibrations) are asymmetric about the principal axis, causing pear-shaped distortions of the nucleus.

The most important vibrational order is the quadrupole vibration $\lambda = 2$. With a proper choice of the coordinate axes, it is possible to describe the nuclear shape by two parameters, β and γ rather than by the five $\alpha_{2\mu}$s. The degree of distortion of the nucleus from a sphere is associated with β. When distorted, considering the nucleus to be formed by rotating an ellipse about the principal axis, β is given by

$$\beta = \Delta R/R_o \qquad (13)$$

where ΔR is the difference between the nuclear radii in the direction of the principal axis and perpendicular to this direction and R_o is the mean nuclear radius. For positive β and with γ fixed at zero, the nucleus is stretched at the poles like a cigar (prolate deformation) and for negative β is flattened at the poles like a saucer (oblate deformation). The parameter γ is related to the shape of the nuclear cross section perpendicular to the principal axis, being a circle for $\gamma = 0$ and an ellipse for other values.

1. Even-Even Nuclides

The theoretical energy level scheme of an even-even nucleus performing quadrupole vibrations is shown in Figure 8. All states have positive parity and the energy difference between each level is constant. The first excited vibrational state (1 phonon) has energy $h\nu_2$ and spin/parity of 2^+. There will be three two-phonon states, all of energy $2h\nu_2$, with spin/parity of 0^+, 2^+, and 4^+. Three-phonon states for quadrupole vibrations will be five in number, of energy $3h\nu_2$, and having spin/parity of 0^+, 2^+, 3^+, 4^+ and 6^+. Because nuclei near closed shells resist deformations, the energy separation decreases as one moves away from magic numbers. In practice, it will not be surprising to find that states which should have exactly the same energy are actually separated because of various interactions which we have not taken into account.

The phonon of the octupole $\lambda = 3$ vibration carries three units of angular momentum and odd parity, resulting in a first excited state of 3^-. Its energy in theory should be approximately that of the two-phonon quadrupole states.

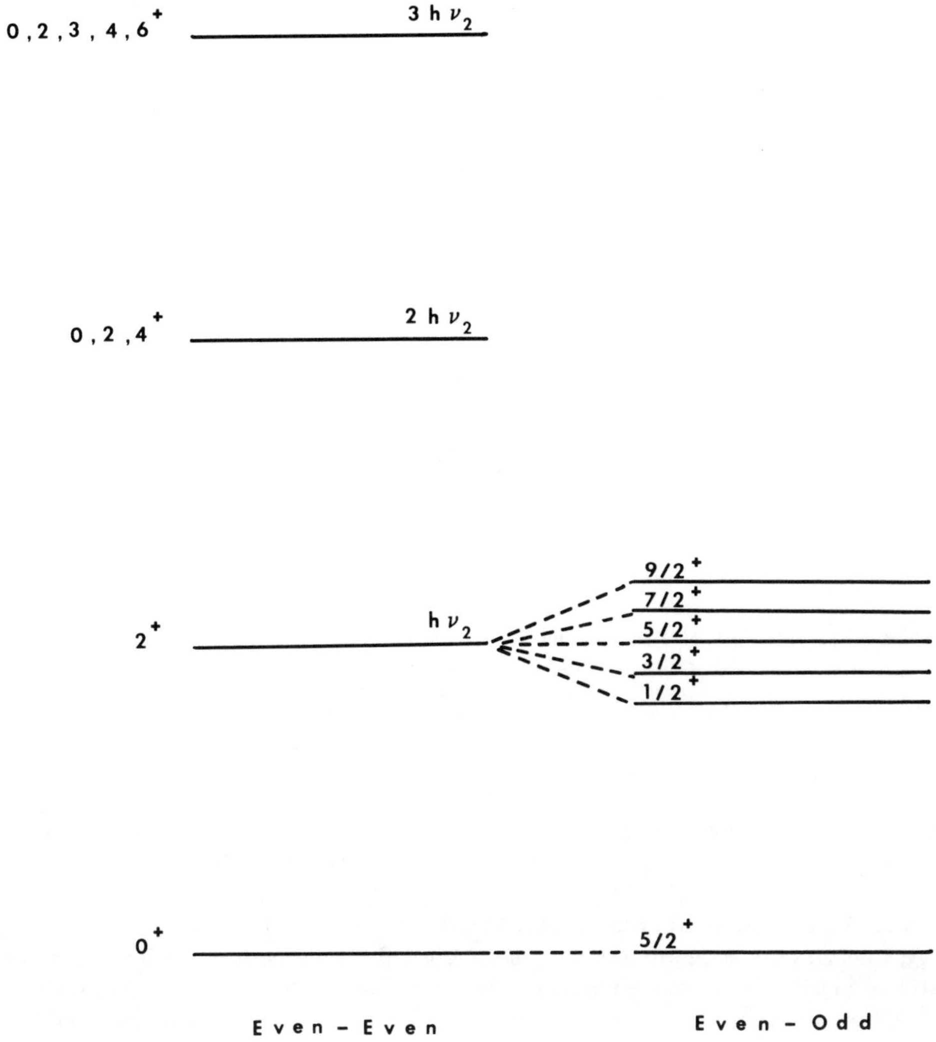

$0,2,3,4,6^+$ _____ $3\,h\,\nu_2$

$0,2,4^+$ _____ $2\,h\,\nu_2$

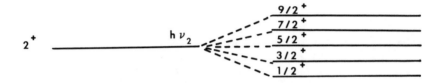

2^+ _____ $h\,\nu_2$

$9/2^+$ _____
$7/2^+$ _____
$5/2^+$ _____
$3/2^+$ _____
$1/2^+$ _____

0^+ _____ _ _ _ _ _ _ _ $5/2^+$ _____

E v e n – E v e n   **E v e n – O d d**

FIGURE 7.   Quadrupole vibrational energy levels of an even-even nucleus (left) and the idealized energy spectrum resulting from the weak coupling of these vibrations to a 5/2⁺ single particle of the neighboring odd-even nucleus (right). Splitting of the second and third excited states is too complex to be shown.

2. Odd-Even Nuclides

In nearly spherical odd-even nuclei, the orbit of the unpaired nucleon will depend to some extent on the form of the nuclear surface. This dependence leads to an interaction of the single particle with vibrations of the core. Since the surface vibrations of nearly spherical nuclei are small, we expect the interaction between single particle and collective motions to be weak. Excited levels of odd-even nuclei then consist of pure single particle states, pure vibrational states, and states produced by weak coupling between the two.

Coupling of a single particle to quadrupole vibrations is generally of greatest interest. If a single particle of angular momentum j is coupled to a one-phonon quadrupole vibration carrying two units of angular momentum, states may be produced with spin ranging from | j − 2 |, to (j + 2) in integral steps. The parity is that of the single particle state. An idealized case is presented in Figure 7 where the 2⁺ one-phonon quadrupole vibrational state of the even-even core couples with a 5/2⁺ single particle ground

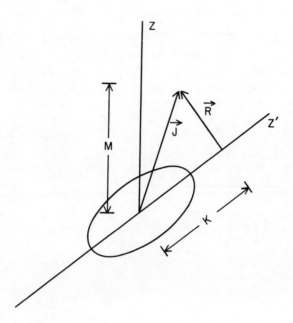

FIGURE 8.   Angular momentum relationships in a permanently deformed nucleus. The total angular momentum J of the nucleus has component M along the z-axis fixed in space and component K along the nuclear symmetry axis z'. The rotational angular momentum R of the nucleus is perpendicular to the z'-axis.

state to produce five excited states. In reality the situation is rarely so clear since states due to other phonon-single particle interactions often appear.

C. Nuclides Far Removed From Doubly Magic

As nucleons are added to a doubly magic core, the nonspherical distribution of the extracore nucleons causes the nucleus to become increasingly susceptible to distortion. When both the number of protons and neutrons are far removed from magic numbers, the nuclear shape becomes permanently spheroidal. Permanent distortion is encountered in nuclides in the neighborhood of A = 25, A = 150 − 190, and A ⩾ 220. By a spheroidal deformation is meant a shape which can be generated by rotating an ellipse either about its major or minor axis. In the former case, a prolate (cigar-shaped) spheroid is obtained while in the latter case an oblate (saucer-shaped) spheroid results. A cross section perpendicular to the axis of revolution is a circle. Prolate deformations for nuclei are favored because it happens that then the spin-orbit interaction acts to lower the energy of certain nucleons. It is the permanent deformation of these nuclei that accounts for their large electric quadrupole moments (Section II.F).

For simplicity let us consider a single nucleon moving within a spheroidal nucleus. We may think of the remainder of the nucleons as comprising a spheroidal core capable of rotating or vibrating. It may be recalled that the angular momentum of an object about a point never changes if its potential energy depends only on the distance of the object from the point. In the spheroidal nucleus, however, the potential energy of the nucleon depends on angle as well as distance so that its angular momentum is not conserved. Since the total angular momentum of the nucleus is always fixed, it follows that the angular momentum of the core must constantly change in coordination with that of the single nucleon so that their sum remains constant.

To investigate further the angular momentum relationships of particle and core, we construct two sets of coordinate axes. One, the x, y, z-system, is fixed in space while the x′, y′, z′-system is fixed in the nucleus. We take z′ to lie along the axis of symmetry of the spheroidal nucleus, as shown in Figure 8. We call J the total angular momentum of the nucleus and j the angular momentum of the outer nucleon. The angular momentum of the core, **R**, may be due to rotational or vibrational motions. Then,

$$J = j + R$$

The component of J along the space-fixed axis z is called M and along z′ is called K. Both M and K have fixed values.

Now, in a semiclassical picture we can imagine the interrelationship of j and **R** as resulting from a particle's bouncing within a spheroidal shell. In these collisions the nucleon imparts angular momentum to the shell, for example, by causing it to rotate. In turn, the moving shell can transfer angular momentum to the particle. However, an exchange of angular momentum between particle and shell with a component along the z′ axis cannot take place. Just as a particle bouncing within a sphere cannot cause the sphere to rotate, so the symmetry of the shell about the z′ axis prohibits the particle from setting the shell into rotation about this axis. Consequently, the component of j along the z′ axis, which we call Ω, remains fixed. If only rotational motions of the shell are considered, the component of **R** along the z′ axis, will be zero, i.e., **R** will be perpendicular to the z′ axis. In that case, the z′ components of J and j will be equal:

$$K = \Omega$$

Because of pairing forces, the example of a single nucleon moving in a deformed nucleus is not as oversimplified as might be thought. Pairing forces cause nucleons beyond the doubly magic core to orient their angular momenta in the ground state so that the Ω of one is equal and opposite to the Ω of the other. Hence, the z′ components of the angular momenta of the pairs vanish so that for even-even nuclei in their lowest state Ω = 0 = K. For odd-even nuclei, only the z′ component of angular momentum of the unpaired nucleon contributes to Ω.

In general, the energy of permanently deformed nuclei is due to three sources: rotational motion, vibrations, and single particle excitations. Fortunately, in most cases the energy associated with each of these motions is different enough so that each can be treated separately. Rotational excitations are generally lowest in energy, followed in order by single particle excitations and vibrational excitations. In even-even nuclei, the large energy required to split a pair of nucleons can raise single particle excitations above the vibrational.

The energy levels associated with each of these motions will now be considered.

1. Rotational Excitations

In classical mechanics the energy due to a rigid object rotating about an axis is

$$E = \frac{L^2}{2I} \tag{14}$$

where L is the angular momentum of the object about the axis and I is the so-called moment of inertia:

$$I = \sum m_i r_i^2$$

where r_i is the distance from the axis to a mass element m_i of the object.

In the quantum mechanical treatment of the rotation of a permanently deformed nucleus, we replace L^2 of Equation 14 by $J(J+1)\hbar^2$ where J is the spin of the nucleus:

$$E_{Rot} = \frac{\hbar^2}{2I} J(J+1) + E_o \qquad K \neq \tfrac{1}{2} \qquad\qquad (15)$$

In this equation which is valid when K does not equal one half, E_o and I are constants. The angular momentum quantum number J can take on a series of values, giving rise to a *rotational band* of energy levels "built on" a particular energy level (the *base state*) of spin J_o and energy E_o. In general,

$$J = J_0, J_0 + 1, J_0 + 2, \cdots$$

In the special case of K = 0, certain symmetry requirements restrict the possible values of J to

$$J = J_0, J_0 + 2, J_0 + 4, \cdots \qquad K = 0$$

The parity of the rotational levels is that of the base state. In the special case of K = ½, a correction term must be added to the energy

$$E_{Rot} = \frac{\hbar^2}{2I} [J(J+1) + a(-1)^{J+1/2}(J+\tfrac{1}{2})] + E_0, \qquad K = \tfrac{1}{2}$$
$$(16)$$

where a is the so-called decoupling constant.

For even-even nuclei in their ground state, K = 0, $E_o = 0$, and E_{Rot} reduces to

$$E_{Rot} = \frac{\hbar^2}{2I} J(J+1)$$

In this especially simple case, nucleons are symmetrically located about a plane through the center of mass and perpendicular to z'. Because of parity considerations, the angular momentum values are restricted to

$$J = 0, 2, 4, \cdots$$

each state having even parity. Deformed even-even nuclei are expected then to have a series of energy levels built on the ground state with spacings

$$E_{4^+}/E_{2^+} = \frac{20}{6} = 3.33$$

$$E_{6^+}/E_{2^+} = \frac{42}{6} = 7$$

$$E_{8^+}/E_{2^+} = \frac{72}{6} = 12$$

Numerous examples of these rotational bands are found and there is usually excellent agreement with theory.

Odd-even nuclei in the ground state generally have

$$J = K = \Omega$$

An exception can occur when $K = \frac{1}{2}$ and the decoupling parameter a in Equation 16 is negative, in which case the ground state can have spin 3/2. In any event, the rotational band will have energies given by Equation 15 or 16 with

$$J = K, K + 1, K + 2, \cdots$$

The parity of the energy levels will be that of the ground state.

Rotational bands may be built on excited single particle states or on vibrational states. The resultant energy spectrum becomes quite complex when, for example, there are several single particle states lying close to each other with intermingled rotational bands.

2. Vibrational Excitations

It will be recalled from Section IV.B that the lowest vibrational order of interest is the quadrupole ($\lambda = 2$). Quadrupole vibrations are of two types: β vibrations in which there are oscillations with preservation of a circular cross section perpendicular to the z' axis, and γ vibrations in which there are departures from a circular cross section. Because they have no angular momentum about the z' axis, β vibrations produce states with $K = 0$ and angular momentum and parity of 0^+, 2^+, 4^+, etc. For γ vibrations, states have $K = 2$ and angular momentum/parity of 2^+, 3^+, 4^+, etc.

For even-even nuclei the lowest vibrational states built upon the ground state will be 0^+ (β vibration) or 2^+ (γ vibration). Commonly, rotational bands may be identified which are built upon one or more of the vibrational states. Octupole ($\lambda = 3$) vibrations may also produce states in the low energy spectrum of even-even nuclides. The one phonon/octupole excitation carries three units of angular momentum with

$$K_V = 0, 1, 2, 3$$

The octupole vibrational angular momentum may then combine with rotational angular momentum having $K_R = 0$ to give a resultant $K = K_V$. Then possible values of J are

$$J = K_V, K_V + 1, K_V + 2, \cdots$$

However, if $K_V = 0$, symmetry requirements limit negative parity states to

$$J = 1^-, 3^-, 5^-, \cdots$$

Levels of 1^- and rotational bands built upon them have indeed been identified in certain nuclides.

Vibrational levels in distorted odd-even nuclides are usually difficult to identify because of the presence of many single particle levels.

3. Single-Particle Excitations

The low lying energy levels of permanently deformed nuclei will include in addition to rotational and vibrational excitations, single particle states similar to those found near the doubly magic nuclei (Section IV.A). In fact the calculation of the order and properties of the single-particle states in deformed nuclei proceeds in much the same manner as that of the single particle model. However, instead of using a harmonic oscillator potential with spherical symmetry, the "spring constant" of the oscillator

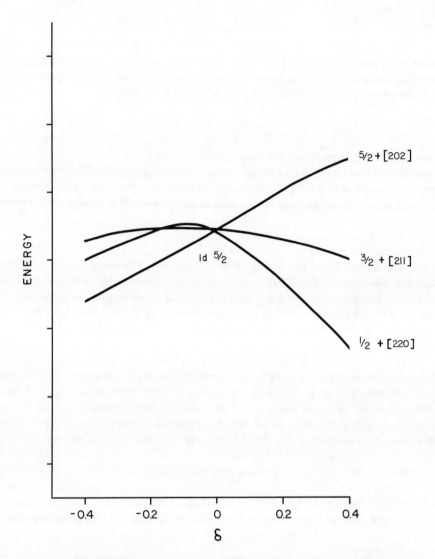

FIGURE 9.   The splitting of the 1d 5/2 energy level due to a spheroidal deformation.
For $\delta = 0$ the nucleus is spherical, while positive and negative values of δ correspond
to prolate and oblate deformations, respectively. (After Nilsson, S. G., *Dan. Mat. Fys.*
Medd., 29(16), 1955. With permission.)

along the z' axis is taken to be different than that along the x' and y' axes. The spin-
orbit and $\boldsymbol{l} \cdot \boldsymbol{l}$ interactions are retained as in Equation 11.

In the single particle model for spherical nuclei the energies of nucleons having the
same $\boldsymbol{l}$ and j were identical. For a nucleus deformed in a prolate shape, these energy
levels are no longer equal, rising when $| \Omega |$, the absolute value of the component of $\boldsymbol{j}$
on the z' axis, is large and falling when $| \Omega |$ is small. The values that Ω may assume
for a particular value of j are, as usual,

$$\Omega = -j, -j + 1, \cdots, j$$

States of $\pm\Omega$ have the same energy so that a pair of neutrons or protons go into each
energy level, one with component Ω and the other with component $-\Omega$.

The greater the nuclear deformation, the greater is the divergence among energy levels of different Ω. In addition, the vector $\boldsymbol{l}$ and j no longer remain stationary but begin to precess rapidly about the z' axis so that the values of $\boldsymbol{l}$ and j change with time. It then becomes necessary to find other quantum numbers which are constants of the motion suitable for identifying the different particle states. For strongly deformed nuclei, Ω remains a good quantum number. Also, the parity of each state, given as before by $(-1)^N$ where N is defined by Equation 10, continues to have a fixed value as does N. Two new quantum numbers appropriate for strongly deformed nuclei are n_z' and Λ. The quantity n_z' is the number of oscillator quanta in the z' direction and Λ is the z' component of $\boldsymbol{l}$. Customarily, each state is labeled by $|\Omega|$, parity, and in brackets by $[N, n_z' \Lambda]$. Thus the state $5/2 + [642]$ has $\Omega = 5/2$, even parity, N = 6, $n_z' = 4$, and $\Lambda = 2$. Since $j = \boldsymbol{l} + s$, where s is the intrinsic spin of the nucleon, the z' components of these vectors are related by $\Omega = \Lambda \pm \frac{1}{2}$. In the above example, s would point in the same direction as $\boldsymbol{l}$.

To summarize, for spherical nuclei energy levels are specified by the quantum numbers N, $\boldsymbol{l}$, and j with each level containing a maximum of $(2j + 1)$ identical nucleons. Deformed nuclei have energy levels specified by N, Ω, n_z', and Λ with each level containing at most two identical nucleons. Figure 9 illustrates how the spacing between energy levels increases with the nuclear deformation δ in a typical case, the $1d_{5/2}$ state. The deformation parameter has essentially the same value as β defined in Equation 13. Similar graphs for all energy levels have been published by Nilsson[8] and are referred to as *Nilsson diagrams*.

a. Odd-Even Nuclides

We expect the pair of identical nucleons which occupy each state to have Ω quantum numbers of equal magnitude but opposite sign. Unless pairs of nucleons have been broken or there are vibrational or rotational excitations, the angular momentum of the nucleus is the angular momentum of the unpaired nucleon. In this case the nuclear spin J is equal to the Ω of the last nucleon. To find the ground state spin and parity for a particular deformed odd-even nucleus, one first finds the nuclear deformation from tabulations.[9] Typically δ lies in the range from 0.2 to 0.3. Then one determines from a Nilsson diagram of energy levels the Ω and parity of the unpaired nucleon according to the ordering of the energy levels for the given δ. This result is the ground state spin and parity of the nucleus except when several levels lie so close together that their precise order is unknown.

The low-lying excited single particle states will consist of the immediately higher levels in the Nilsson diagram, assuming that the energy required to split up pairs is relatively large. Rotational bands built on the ground and excited single particle states are to be expected.

b. Odd-Odd Nuclides

The angular momenta of the two unpaired nucleons are expected to couple so that two nuclear spins are possible:

$$J = |\Omega_p \pm \Omega_n|$$

where Ω_p and Ω_n are the z' components of the angular momenta of the unpaired proton and neutron, respectively. Rules somewhat analogous to Nordheim's rules for spherical odd-odd nuclei have been advanced by Gallagher and Moszkowski for predicting the spin of the lower of the two levels. The rule states that one must look at the relation $\Omega = \Lambda \pm \frac{1}{2}$ for the proton and neutron separately:

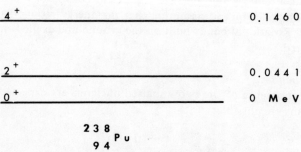

———————————————————— (4$^+$)   1.1258

————————————————————   3$^+$ 1.0699

————————————————————   2$^+$ 1.0285

———————————————————— 2$^+$ 0.9830

———————————————————— 0$^+$ 0.9415

———————————————————— 5$^-$ 0.7632

———————————————————— 3$^-$ 0.6614

———————————————————— 1$^-$ 0.6051

8$^+$ ————————————————————   0.514

6$^+$ ————————————————————   0.3034

4$^+$ ————————————————————   0.1460

2$^+$ ————————————————————   0.0441

0$^+$ ————————————————————   0 MeV

$^{238}_{94}$Pu

FIGURE 10.   Energy levels of the permanently deformed, even-even $^{238}_{94}$Pu nucleus. Spin/parity assignments are vertically aligned to display rotational bands built upon the ground and three vibrational states.

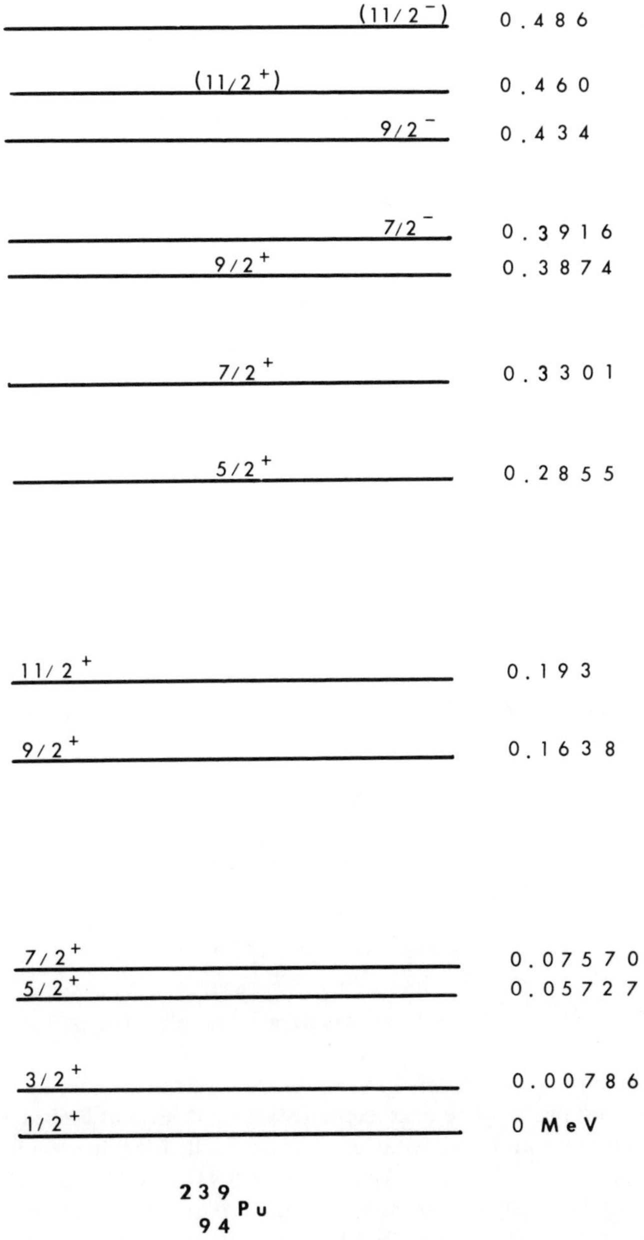

FIGURE 11. Energy levels of the permanently deformed, odd-even $^{239}_{94}$Pu nucleus. Spin/parity assignments are vertically aligned to display three rotational bands built upon single particle states.

1.   If Ω_p is related to Λ_p by the same sign of $\frac{1}{2}$ as Ω_n is related to Λ_n, then the spin of the lower energy state is

$$J = \Omega_p + \Omega_n$$

2.   If Ω_p is related to Λ_p by the opposite sign of $\frac{1}{2}$ as Ω_n is related to Λ_n, then the spin of the lower energy state is

$$J = |\Omega_p - \Omega_n|$$

The theoretical basis for the Gallagher-Moszkowski rule is that there exists an interaction between the odd proton and neutron which lowers their energy when their intrinsic spin directions are parallel.

c. Even-Even Nuclides

Single particle excitations in even-even nuclei are likely to be of relatively high energy because of the energy required to break a pair of identical nucleons. If a pair is broken and one member moved to a higher state, its new angular momentum may be recoupled with that of its original partner. Then if Ω_1 and Ω_2 are the z' components of their angular momenta, the coupled angular momentum can have either of two z' components:

$$\Omega = |\Omega_1 \pm \Omega_2|$$

This situation is reminiscent of the odd-odd nucleus. But whereas the interaction between proton and neutron lowers the energy when their intrinsic spins are parallel, identical nucleons have the lowest energy when their spins are antiparallel. Thus Gallagher has proposed a rule stating that in examining the relation $\Omega = \Lambda \pm \frac{1}{2}$ for each of the two particles:

1.   If Ω_1 is related to Λ_1 by the same sign of $\frac{1}{2}$ as Ω_1 is related to Λ_2, then the spin of the lower level is

$$J = |\Omega_1 - \Omega_2|$$

2.   If Ω_1 is related to Λ_1 by the opposite sign of $\frac{1}{2}$ as Ω_2 is related to Λ_2, then the spin of the lower level is

$$J = \Omega_1 + \Omega_2$$

Example 4 — $^{238}_{94}$Pu. Pu-238 is an excellent example of a permanently deformed, even Z-even N nucleus. Its energy level scheme is shown in Figure 10. The first four excited states are rotational levels built upon the 0^+ ground state, having spins of 2, 4, 6, and 8 with even parity. The next excited state, a 1^- level at 0.6051 MeV, is an octupole-vibrational state and two rotational levels built upon it are seen at 0.6614 and 0.7632 MeV. A 0^+, K = 0 β-vibrational state at 0.9415 MeV and a 2^+, K = 2 γ-vibrational state at 1.0285 MeV also are seen with their associated rotational bands.

Example 5 — $^{239}_{94}$Pu. The even Z-odd N nucleus of ^{239}Pu is typical of many odd-A deformed nuclei. Its low-lying levels consist of rotational bands built upon three single particle states. With 145 neutrons the ground state of ^{239}Pu is the $\frac{1}{2}$ + [631] state. The first five excited levels are built upon this state and have the irregular spacings characteristic of K = $\frac{1}{2}$ bands. The levels at 0.3301, 0.3874, and 0.460 MeV are members of a rotational band built on the 5/2 + [622] state at 0.2855 MeV. The third base state is a 7/2 − [743] level at 0.3916 MeV on which are based the $9/2^-$ and $11/2^-$ states at 0.434 and 0.486 MeV, respectively. The level scheme of ^{239}Pu may be seen in Figure 11.

REFERENCES

1. Mayer, M. G., Nuclear configurations in the spin orbit coupling model. I. Empirical evidence, *Phys. Rev.*, 78, 16, 1950.
2. Haxel, O., Jensen, J. H. D., and Suess, H. E., Modellmassige deutung der ausgezeichneten nukleonen-zahlen im kernbau, *Z. Physik*, 128, 295, 1950.
3. True, W. W., Ma, C. W., and Pinkston, W. T., Negative-parity states in Pb^{208}, *Phys. Rev.*, C3, 2421, 1971.
4. Dunnweber, W., Cosman, E. R., Grosse, E., Hering, W. R., and Von Brentano, P., Gamma decay of particle-core states in Pb^{209}, *Nucl. Phys.*, A247, 251, 1975.
5. Redlich, M. G., Energy levels of Pb^{210} from shell theory, *Phys. Rev.*, 138 (3B), B544, 1965.
6. Bohr, A. and Mottleson, B. R., Collective motion and nuclear spectra, in *Nuclear Spectroscopy, Part B*, Ajzenberg-Selove, F., Ed., Academic Press, New York, 1960.
7. Cohen, B. L., *Concepts of Nuclear Physics*, McGraw-Hill, New York, 1971.
8. Nilsson, S. G., Binding states of individual nucleons in strongly deformed nuclei, *Dan. Mat. Fys. Medd.*, 29 (16), 1955.
9. Mottleson, B. R. and Nilsson, S. G., The intrinsic states of odd-A nuclei having ellipsoidal equilibrium shape, *Mat. Fys. Skr. Dan. Vid. Selsk.*, 1 (8), 1959.

Chapter 2

MODES OF NUCLEAR DECAY

Donald A. Walker

TABLE OF CONTENTS

I.   Natural Radioactivity...38

II.   The Radioactive Decay Curve..38

III.   Simple Systematics of Radioactive Decay 39
        A.   Alpha Decay 40
        B.   Beta Decay..41
        C.   Gamma Decay...43
        D.   Internal Conversion 46

IV.   Binding Energy 47

V.   A Detailed Analysis of Alpha Decay 48
        A.   Alpha Decay Energetics....................................48
        B.   Alpha Particle Spectra....................................51
        C.   Energy and Half-Life......................................52
        D.   Theory of Alpha Decay 52

VI.   A Detailed Analysis of Beta Decay...................................59
        A.   Beta Particle Spectra 59
        B.   Beta Decay Energetics 60
        C.   Fermi Theory of Beta Decay 63
        D.   Ratio of Electron Capture to Positron Emission 67

VII.   A Detailed Analysis of Gamma Decay..................................68
        A.   Gamma Ray Spectra...68
        B.   The Theory of Gamma Ray Emission 68
        C.   Internal Conversion Coefficients 76

References 78

I. NATURAL RADIOACTIVITY

The discovery of natural radioactivity in 1896 by Henri Becquerel marked the beginning of atomic and nuclear physics. Working initially with uranium, Becquerel discovered that certain naturally occurring substances emitted radiations that could expose photographic plates and discharge electrically charged bodies.

Further investigations by Rutherford, the Curies, and others showed that many other naturally occurring substances emitted radiations. In time it was determined that there were three types of radiations that accounted for the phenomenon that was called radioactivity. The three types of radiations were named:

1. Alpha particles (α)
2. Beta particles (β)
3. Gamma rays (γ)

The substances that emit one or more of these radiations are called radioactive nuclides.

Investigations since the early years of this century have established that the three types of radiations originate in the nucleus of certain nuclides, and that (1) alpha particles (α) are, in actuality, helium nuclei ($^4_2$He); (2) beta particles (β^- and β^+) are, in actuality, negative and positive electrons (e^- and e^+); and (3) gamma rays (γ) are high energy electromagnetic radiations (photons).

Radioactive disintegration is a mechanism whereby an unstable nucleus can give up energy in order to achieve a configuration of greater stability.

II. THE RADIOACTIVE DECAY CURVE

If N represents the number of radioactive nuclei in a certain sample at a certain time, then the rate at which the nuclei in the sample decay is given by

$$\frac{-dN}{dt}$$

where the minus sign signifies that the number of nuclei in the sample decreases with time.

The quantity $-(dN/dt)$ is called the *activity* of the sample:

$$\text{ACTIVITY} = \frac{-dN}{dt}$$

The activity of a radioactive substance is often measured in terms of the *curie*, which is defined as follows:

$$1 \text{ curie (Ci)} = 3.70 \times 10^{10} \text{ disintegrations per second}$$

$$(1 \text{ mCi} = 10^{-3} \text{ Ci})$$

$$(1 \text{ } \mu\text{Ci} = 10^{-6} \text{ Ci})$$

Activity is sometimes measured in terms of the *rutherford*, which is defined as follows:

$$1 \text{ rutherford (rd)} = 10^6 \text{ disintegrations per second}$$

$$(1 \text{ mrd} = 10^{-3} \text{ rd} = 10^3 \text{ disintegrations per second})$$

$$(1 \text{ } \mu\text{rd} = 10^{-6} \text{ rd} = 1 \text{ disintegration per second})$$

The disintegration of radioactive nuclei is a random process. However, when a large number of nuclei are considered it is found that the activity of a radioactive substance is directly proportional to the number of radioactive atoms present, i.e.:

$$\frac{-dN}{dt} = \lambda N$$

where λ is a proportionality constant called the *radioactive decay constant*. It has dimensions of reciprocal time, and has a characteristic value for each radioactive nuclide. Arranging the above differential equation as follows:

$$\frac{dN}{N} = -\lambda \, dt$$

we can integrate to get $\ln N - \ln N_0 = -\lambda t$ where N_0 is the number of radioactive nuclei present at $t = 0$. Then $\ln (N/N_0) = -\lambda t$, and, exponentiating, we get

$$N(t) = N_0 \, e^{-\lambda t}$$

This result shows that the number of radioactive nuclei present at time t decreases exponentially.

The *half-life* of a radioactive substance is defined as the length of time required for half of the original nuclei to decay. It is usually symbolized by $T_{1/2}$. Using this definition we can write the previous equation for the case where $t = T_{1/2}$, and $N(T_{1/2}) = N_0/2$. Then

$$\frac{N_0}{2} = N_0 \, e^{-\lambda T_{1/2}}$$

Therefore, $\lambda T_{1/2} = \ln 2 = 0.69315$.

This is the basic relationship between the half-life of a substance and its decay constant.

III. SIMPLE SYSTEMATICS OF RADIOACTIVE DECAY

Atomic nuclei are often designated as:

$$^A_Z X$$

where X represents the chemical symbol for the substance. A (sometimes called the *mass number*) represents the total number of heavy particles (protons and neutrons — referred to collectively as *nucleons*) in the nucleus. A is often referred to as the *nucleon number*.

Z is usually referred to as the *atomic number*. Since it represents the number of protons in the nucleus, it is also referred to as the *proton number*.

Sometimes there is occasion to refer to the *neutron number*, N, for a nucleus. In terms of A and Z, $N = A - Z$.

The term *nuclide* which we have been using refers to a specific nucleus with atomic number Z and mass number A. This is what we have symbolized by ^A_ZX. Nuclides with the same Z are called *isotopes*; nuclides with the same A are called *isobars*; nuclides with the same N are called *isotones*. Nuclides that have the same A and Z but are in different states of excitation are called *isomers*.

Example — There are three stable isotopes of oxygen. The atomic number of oxygen is 8 (Z = 8); the mass numbers for the three isotopes are 16, 17, and 18. Hence the three isotopes would be represented symbolically as:

$$^{16}_8O, \ ^{17}_8O, \text{ and } ^{18}_8O$$

Each of the three isotopes has 8 protons in its nucleus; they have 8, 9, and 10 neutrons, respectively. Since the atomic number is redundant — inasmuch as the chemical symbol uniquely determines it — the above isotopes might be expressed symbolically as:

$$^{16}O, \ ^{17}O, \text{ and } ^{18}O$$

In alpha and beta decay an isotope of one element undergoes a transformation to an isotope of another element. The initial nucleus is referred to as the *parent nucleus*; the final nucleus the *daughter nucleus*.

Let ^A_ZX represent the parent nucleus.

Let $^{A'}_{Z'}$Y represent the daughter nucleus.

A. Alpha Decay

In terms of the above convention, alpha decay can be represented by the following:

$$^A_Z X \rightarrow \ ^{A-4}_{Z-2} Y + \ ^4_2 He$$

where the emitted alpha particle is shown as a helium nucleus.

Since the atomic number of the daughter nucleus is different from that of the parent, the two nuclei are isotopes of different elements with different chemical properties.

The above transformation also demonstrates two important conservation laws that are general for all radioactive decays:

1.   Conservation of electric charge. There are Z protons before and after the decay.
2.   Conservation of nucleons. There are A nucleons before and after the decay.

As an example of alpha decay we will refer to the alpha decay of the most abundant isotope of uranium, $^{238}_{92}$U:

$$^{238}_{92}U \rightarrow \ ^{234}_{90}Th + \ ^4_2 He$$

The daughter nucleus is thorium-234. Note that the daughter nucleus has four fewer nucleons and two fewer protons than the parent.

Alpha decay can be shown diagrammatically on a *chart of the nuclides*. This is a chart on which all the known nuclides are plotted with the Z number on the vertical axis and the N number on the horizontal axis. A section of the chart of the nuclides which shows the above alpha decay is shown in Figure 1.

All alpha transitions on the chart of the nuclides show Z decreasing by 2 and N decreasing by 2.

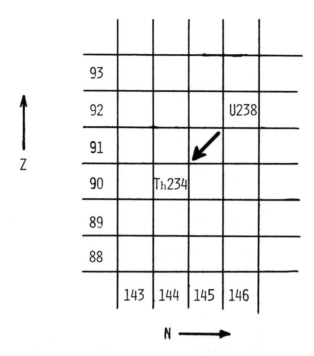

FIGURE 1.   Section of the chart of the nuclides showing the
alpha decay $^{238}_{92}U \rightarrow {}^{234}_{90}Th + {}^{4}_{2}He$.

B. Beta Decay

The term beta decay encompasses three types of nuclear transformations: β^- decay, β^+ decay, and electron capture.

Using the previous symbolic conventions we can represent β^- decay as:

$$^{A}_{Z}X \rightarrow {}^{A}_{Z+1}Y + e^- + \bar{\nu}_e$$

In this process two particles are emitted from the nucleus. One, the *beta particle,* e^-, is a negative electron. The second, the antineutrino, $\bar{\nu}_e$, is an antiparticle of the neutrino and carries no charge. For decades the neutrino (and antineutrino) were considered zero rest mass particles that always accompanied beta decay. However, recent research indicates that neutrinos may possess a small rest mass. Neutrinos and antineutrinos will be discussed further when beta decay is considered in detail later on.

Note in the beta decay transition above that the atomic number of the daughter nucleus is different from that of the parent, i.e., they are isotopes of different elements. Note that the mass number of parent and daughter nuclei is the same. This is always true in beta decay.

In addition to conservation of charge and conservation of nucleons (previously discussed), the beta decay transformation demonstrates another general conservation law: conservation of leptons. Leptons are a class of particles (the so-called light particles) made up of electrons, muons, and neutrinos. (Recent studies indicate that there may be another lepton, a "heavy" lepton, called the tau.)

There are different kinds of neutrinos: the electron neutrino, ν_e, and the muon neutrino, ν_μ (and possibly the tau neutrino, ν_τ). Each of these particles has a corresponding antiparticle. To apply the conservation of leptons law, each particle (electron, muon, tau, electron neutrino, muon neutrino, tau neutrino) is given a *lepton number* of $+1$, whereas the corresponding antiparticles are assigned a lepton number of -1. All other particles are assigned a lepton number of zero. Then the conservation of leptons law states that in a given interaction the total number of leptons is conserved. As an example, the decay of the neutron may be written as:

$$n \rightarrow p + e^- + \nu_e$$

where the lepton number before and after the decay is zero.

Using the conventional symbols we can represent β^+ *decay* as:

$$^A_Z X \rightarrow {}^{A}_{Z-1} Y + e^+ + \nu_e$$

where the emitted beta particle is a positive electron known as a *positron*. Accordingly, this decay is referred to as positron decay. In this decay e^+ is the antiparticle of the electron (lepton number -1) and the neutrino, ν_e, is the particle (lepton number $+1$). Hence we see that conservation of leptons holds here just as it did before.

A distinguishing feature of positron decay is that a positron when emitted will, after losing most of its energy through ionization losses and/or radiation losses, ultimately combine with a negative electron. The two particles will *annihilate*, i.e., the masses of the two particles will be converted into an equivalent amount of radiant energy according to the relativistic expression $E = mc^2$. Therefore, the *annihilation radiation* will consist of photons whose energies total $2m_ec^2 = 1.022$ MeV. In most (but not all) annihilations two photons, each having an energy of 0.511 MeV ($= m_ec^2$) will be emitted 180° apart in order to satisfy the law of conservation of linear momentum (i.e., momentum before annihilation $\simeq 0$; momentum after annihilation $\simeq 0$).

A third type of beta decay is *electron capture* in which an orbiting atomic electron is captured by the nucleus and a neutrino is emitted. According to the quantum theory the wave functions of extranuclear electrons can be nonzero at the nucleus. This means that extranuclear electrons have a finite probability of being in the nucleus and being captured there. Although capture of L-, M-, and higher shell electrons is possible, it turns out that it is the innermost or K-shell electrons that have the greatest probability of being captured.

Symbolically, electron capture would be shown as:

$$^A_Z X + e^- \rightarrow {}^{A}_{Z-1} Y + \nu_e$$

where the e^- represents the orbital electron that is captured by the parent nucleus. Following capture, there is a vacancy left in the electron shell which when filled by another atomic electron gives rise to characteristic X-rays or Auger electrons. Electron capture is a decay mode that usually competes with β^+ emission.

The three types of beta decays described above illustrate a basic property of nucleons: their ability to change from one type to another. In simplest form, then, the various nucleon transitions that accompany the three types of beta decays can be written as follows:

- $n \rightarrow p + e^- + \bar{\nu}$ (β^- decay)
- $p \rightarrow n + e^+ + \nu$ (β^+ decay)
- $p + e^- \rightarrow n + \nu$ (electron capture)

Hence we see that β^- decay is, essentially, the transformation of a neutron into a proton; β^+ decay the transformation of a proton into a neutron; and electron capture the combining of an extranuclear electron with a proton to form a neutron.

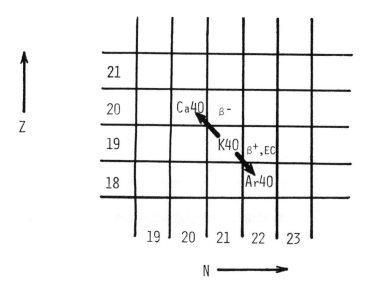

FIGURE 2. Section of the chart of the nuclides showing the beta decay of $^{40}_{19}$K.

The decay of $^{40}_{19}$K provides specific examples of all three beta decay processes:

- $^{40}_{19}$K $\rightarrow$ $^{40}_{20}$Ca + e⁻ + ν (β^- decay)
- $^{40}_{19}$K $\rightarrow$ $^{40}_{18}$Ar + e⁺ + ν (β^+ decay)
- $^{40}_{19}$K + e⁻ $\rightarrow$ $^{40}_{18}$Ar + ν (electron capture)

It should be pointed out that the fact that $^{40}_{19}$K decays by all three beta decay modes is rather unusual. While many β^+ emitters also decay by electron capture (competing modes of decay), it is somewhat uncommon to have a nuclide decay by all three modes.

Beta decay can be shown diagrammatically on the chart of the nuclides. The decay of $^{40}_{19}$K is shown in Figure 2.

The chart shows Z increasing by 1 and N decreasing by 1 (neutron transformed into a proton) for the β^- transition; Z decreasing by 1 and N increasing by one (proton transformed into a neutron) for the β^+ and electron capture transitions.

C. Gamma Decay

Nuclei are ordinarily in a state of minimum energy — their ground state. Under certain circumstances, however, nuclei may be in excited states — quantized, allowed states which are higher in energy than the ground state. In alpha and beta decay we have seen that some nuclei lose excess energy by means of particle emission. When particle emission is not energetically possible, deexcitation of a nucleus can take place via the emission of an electromagnetic wave. A photon emitted from a nucleus during deexcitation is called a gamma ray. (The distinction between X-rays and gamma rays is that X-rays originate outside the nucleus as a result of changes in atomic electron states; gamma rays originate from the nucleus.) How do nuclei reach excited states? They may be excited by photon or particle bombardment, or, as a result of alpha or beta decay, daughter nuclei may be left in excited states from which they decay by means of gamma emission. In any case, when a gamma ray is emitted by a nucleus which moves from an excited state to a state of lower energy the emitted gamma ray photon will have an energy equal to the difference in energy between the initial and final states. Thus:

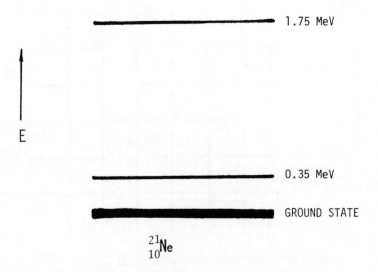

FIGURE 3.   Energy level diagram for $^{21}_{10}$Ne.

$$E\gamma = \Delta E = E_i - E_f = h\nu$$

where E_γ is the energy of the gamma ray, E_i the energy of the initial nuclear state and E_f the energy of the final nuclear state. Using our previous notation, a gamma ray transition to the ground state can be written as

$$(^A_Z X)^* \to {}^A_Z X + \gamma$$

where $(^A_Z X)^*$ represents the excited nucleus.

The above statements can be described pictorially in an energy level diagram such as the one shown in Figure 3 for $^{21}_{10}$Ne.

The vertical axis in an energy level diagram represents energy. The ground state is taken as zero energy and the energy of allowed, quantized excited states (usually measured in MeV) is reckoned with respect to the ground state. For $^{21}_{10}$Ne we see that the first two allowed excited states above the ground state are at energies of 0.35 MeV and 1.75 MeV. If the $^{21}_{10}$Ne nucleus is excited to 1.75 MeV above the ground state (as it is in the β^- decay of $^{21}_9$F), the excess energy of excitation can be released via gamma ray emission as shown in Figure 4.

The diagram shows that the decay from the 1.75 MeV second excited state can go by two modes:

1.   A direct transition to the ground state with the release of a 1.75 MeV gamma ray
2.   A cascade consisting of two transitions where two gamma rays are emitted with energies $E_1 = (1.75 - 0.35) = 1.40$ MeV and $E_2 = 0.35$ MeV

In the above discussion we have said that the energy of the emitted gamma ray is *exactly* equal to the difference in energy between the initial and final states:

$$E\gamma = \Delta E = E_i - E_f = h\nu$$

Actually, this is not strictly true inasmuch as the nucleus must recoil when the gamma ray is emitted and so must assume some of the transition energy. If the excited nucleus is at rest initially it must take on the same magnitude of momentum as the gamma ray in order that momentum be conserved. This is shown in Figure 5.

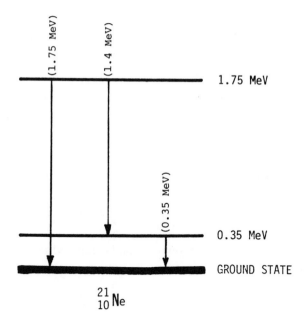

FIGURE 4.   The gamma ray decay of $^{21}_{10}$Ne.

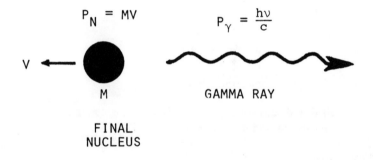

FIGURE 5.   Nucleus of mass M recoiling with velocity V upon the emission of a gamma ray, showing conservation of momentum.

We see that p_N = momentum of final nucleus = MV.

$$p_\gamma = \frac{h\nu}{c}$$

Since

$$p_N = p_\gamma, \quad MV = \frac{h\nu}{c}$$

From energy considerations:

$$E_{Tot} = E_i - E_f = h\nu + \tfrac{1}{2}MV^2$$

$$= h\nu + \frac{(MV)^2}{2M}$$

$$= h\nu + \frac{(h\nu)^2}{2Mc^2}$$

$$(h\nu)^2 + 2Mc^2\,(h\nu) - 2Mc^2\,(E_i - E_f) = 0$$

Solving this quadratically for $h\nu$:

$$h\nu = -Mc^2 + Mc^2 \left\{ 1 + \frac{2(E_i - E_f)}{Mc^2} \right\}^{1/2}$$

Now, since the energy $E_i - E_f$ is extremely small compared to the rest energy of the nucleus, we can approximate:

$$h\nu \simeq -Mc^2 + Mc^2 \left(1 + \frac{E_i - E_f}{Mc^2} \right)$$

$$= -Mc^2 + Mc^2 + E_i - E_f$$

$$= E_i - E_f$$

So, to an excellent approximation we can take:

$$E\gamma = E_i - E_f = \Delta E$$

Accordingly, in all our discussions, we shall ignore the negligible recoil energies involved in the various decays under consideration.

D. Internal Conversion

Internal conversion is a mode of decay which competes with gamma ray emission. In the internal conversion process energy provided in the deexcitation of a nucleus, $E_1 - E_2$, is transferred directly to an extranuclear electron which is then ejected from the atom with a kinetic energy E_K. Such ejected electrons, which can readily be detected, are called conversion electrons. The kinetic energy of conversion electrons is equal to the transition energy ΔE ($= E_1 - E_2$) minus the binding energy of the electron, E_B:

$$E_K = \Delta E - E_B \text{ (recoil energy ignored)}$$

The binding energy of the conversion electron depends upon the element involved and the orbit from which the electron originates — the K-shell, L-shell, etc.

Following internal conversion another atomic electron fills the vacancy left by the conversion electron, resulting in the production of characterisitc X-rays or Auger electrons.

It should be emphasized that internal conversion is a one-step process where the nuclear excitation energy is transferred directly to the extranuclear electron; it is *not* a two-step process in which a gamma ray emitted by the nucleus is subsequently absorbed by the atomic electron in a kind of internal photoelectric effect.

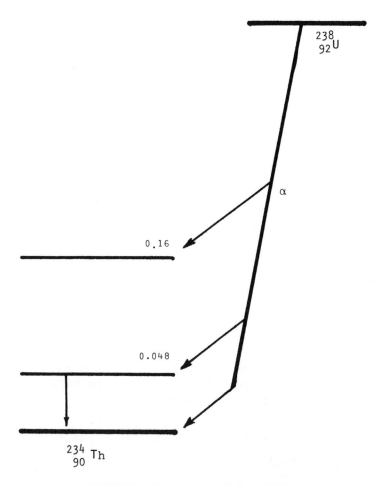

FIGURE 6. The alpha particle decay of $^{238}_{92}$U.

Example — α-Decay. Figure 6 shows the alpha decay of $^{238}_{92}$U:

$$^{238}_{92}\text{U} \rightarrow {}^{234}_{90}\text{Th} + \alpha$$

Note that the ground state and two excited states are populated by the alpha decay. Note also that there is a 0.048 MeV gamma ray emitted as the first excited state of $^{234}_{90}$Th decays to the ground state.

Example — β^--Decay, β^+-decay, and electron capture. Figure 7 shows the decay of $^{40}_{19}$K. As mentioned before, the $^{40}_{19}$K decay is unusual in that it follows the three decay modes β^-, β^+, and electron capture. Note that 89% of the decays of $^{40}_{19}$K are via β^- emission to the ground state of the $^{40}_{20}$Ca daughter. Most of the remaining decays ($\simeq 11\%$) are via electron capture to the first excited state of the $^{40}_{18}$Ar daughter. This state deexcites by means of the emission of a 1.460 MeV gamma ray.

IV. BINDING ENERGY

Before investigating radioactivity in more detail, let us review the concept of binding energy, for this concept will answer the question of when alpha or beta emission can and cannot occur.

Suppose two particles, A and B, are bound by means of some force to form a com-

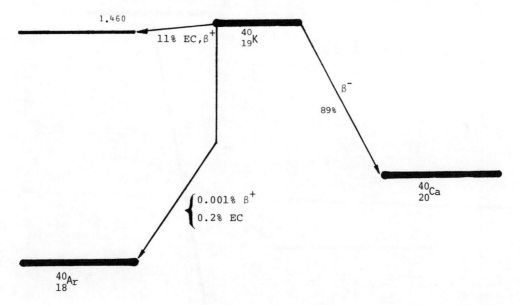

FIGURE 7.   The decay of $^{40}_{19}$K.

posite system with rest mass M_{AB}. We know that it will take a certain energy, E_B, to separate the bound system, i.e., to move the constituent particles A and B an infinite distance from one another. The energy E_B is referred to as the binding energy. Writing the rest masses involved in terms of energy we have:

$$M_{AB}c^2 + E_B = M_A c^2 + M_B c^2$$

The binding energy can then be written as:

$$E_B = (M_A + M_B - M_{AB}) c^2$$

In this form we see that for E_B to be positive, as it must be for a bound system, $(M_A + M_B) - M_{AB}$ must be a positive quantity. Therefore, we see that $M_A + M_B > M_{AB}$. In words, this means that the sum of the masses of the constituent particles must be greater than the mass of the composite system.

Let us consider the implication of a negative E_B. This would mean that $M_{AB} > M_A + M_B$ and so the original system would be unstable.

We shall see that the concept of binding energy is related to the definition of Q values which follows.

V. A DETAILED ANALYSIS OF ALPHA DECAY

A. Alpha Decay Energetics
In the generalized notation introduced previously alpha decay was written as:

$$^A_Z X \rightarrow \ ^{A-4}_{Z-2} Y + \ ^4_2 He$$

If the parent nucleus is assumed to be at rest initially, then conservation of energy in alpha decay would lead to:

$$M_X c^2 = M_Y c^2 + M_\alpha c^2 + K_Y + K_\alpha$$

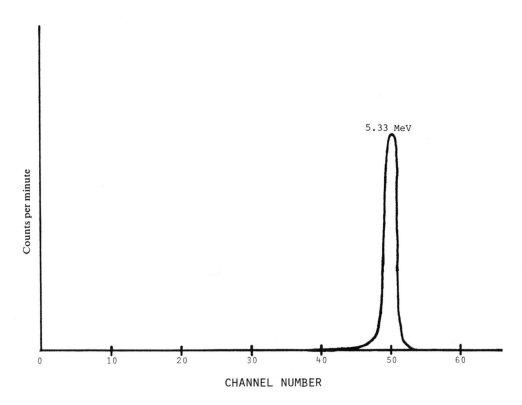

FIGURE 8. Alpha particle spectrum of $^{206}_{84}\text{Po}$.

where K_α represents the kinetic energy of the emitted alpha particle and K_Y the recoil energy of the daughter nucleus. For convenience, all masses in an expression as that above are usually taken as atomic rest masses — electron binding energies have been ignored. The energy released in the decay is called the *disintegration energy* and is symbolized by Q:

$$Q = K_Y + K_\alpha = (M_X - M_Y - M_\alpha)\,c^2$$

Alpha decay is energetically possible only if $Q > 0$. This occurs when $M_X > M_Y + M_\alpha$, i.e., when the mass of the parent nucleus is greater than the mass of the daughter nucleus plus the mass of the alpha particle. Nuclei with A values approximately equal to 144 and larger spontaneously decay by the emission of alpha particles. An analysis similar to the one above shows that mass considerations preclude the spontaneous emission of protons and neutrons — as well as most other particles. It is not uncommon for the alpha particle to be emitted, however, because of its exceedingly large binding energy. (It is doubly magic with $Z = N = 2$, and as a result has a relatively small composite mass in comparison to the sum of the masses of its constituents.)

Momentum must be conserved in an alpha-particle decay. If we assume that $M_X v_X = 0$ before the alpha particle is emitted, then:

$$M_Y v_Y = M_\alpha v_\alpha$$

Therefore:

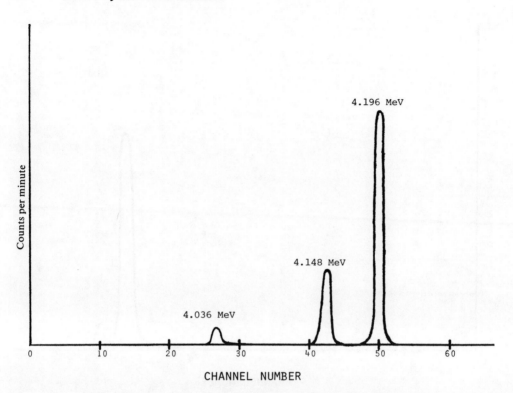

FIGURE 9.   Alpha particle spectrum of $^{238}_{92}$U.

$$Q = K_Y + K_\alpha$$

$$= \tfrac{1}{2} M_Y v^2_Y + K_\alpha$$

$$= \frac{(M_Y v_Y)^2}{2M_Y} + K_\alpha$$

$$= \frac{(M_a v_a)^2}{2M_Y} + K_\alpha$$

$$= \frac{M_\alpha}{M_Y} K_\alpha + K_\alpha$$

$$= K_\alpha \left(1 + \frac{M_\alpha}{M_Y}\right)$$

Then:

$$Q \simeq K_\alpha \left(1 + \frac{4}{A-4}\right) = K_\alpha \left(\frac{A}{A-4}\right)$$

Therefore, with this modest approximation

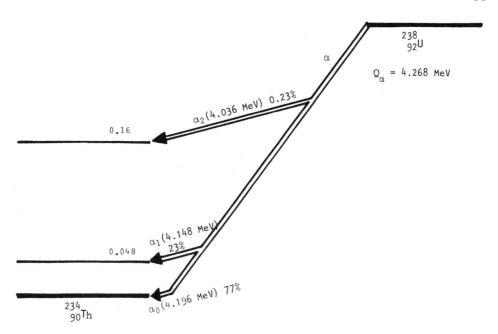

FIGURE 10.   Energy level diagram for the alpha particle decay of $^{238}_{92}$U.

$$K_\alpha = \frac{A-4}{A} \, Q$$

and we see that the alpha particle carries away most of the disintegration energy available in the decay — but not all. Since alpha decay occurs primarily for nuclei of large A, the fraction (A-4)/A is large — usually relatively close to one.

B. Alpha Particle Spectra

If the alpha particles emitted from $^{206}_{84}$Po are detected, the energy spectrum as displayed on a multichannel analyzer would be as shown in Figure 8. The diagram shows that the alpha particles from $^{206}_{84}$Po are essentially monoenergetic.

If the alpha particles emitted from $^{238}_{92}$U are examined with a multichannel analyzer, the spectrum will be as shown in Figure 9.

Figure 9 shows that the alpha particles emitted from $^{238}_{92}$U, while not monoenergetic, are discrete in energy: about 77% of the alphas have an energy of 4.196 MeV, about 23% have an energy of 4.148 MeV, and about 0.23% have an energy of 4.036 MeV. The discrete nature of alpha particle energies is characteristic of all alpha particle emitters and stems from conservation of energy and conservation of momentum considerations which, as shown above, led to the development of the expression

$$K_\alpha \simeq \frac{A-4}{A} \, Q$$

for the kinetic energy of an emitted alpha particle.

The alpha particle spectrum of $^{238}_{92}$U is explained by the energy level diagram shown in Figure 10.

The energy level diagram shows that the parent nucleus decays by alpha emission to either the ground state of $^{234}_{90}$Th or its first or second excited state. The most energetic alpha particle (4.196 MeV) is emitted for the transition to the ground state of the daughter. The next alpha particle has an energy 0.048 MeV smaller than the maximum energy, etc.

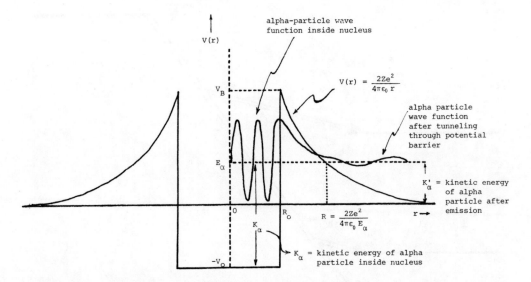

FIGURE 11.   Generalized coulomb potential and simplified nuclear potential for an alpha particle illustrating the tunneling phenomenon.

The multiplicity of lines in alpha particle spectra is usually referred to as fine structure, and such structure has been of great value in studying nuclear energy levels.

C. Energy and Half-Life

In his early investigations, Rutherford commented that some correlation seemed to exist between the half-life of alpha emitters and the range (or energy) of the alpha particles emitted. He noted that the most energetic alpha particles are emitted from nuclei with the shortest half-lives and that the least energetic alpha particles are emitted from nuclei with the longest half-lives.

The theoretical explanation of this observation was not forthcoming until the development of quantum mechanics in the late 1920s. In fact, the successful explanation of alpha particle emission in 1928 by Gamow and independently by Gurney and Condon was one of the first triumphs of the then new quantum mechanics.

Classical physics was unable to explain the alpha decay process since the observed energies of emitted alpha particles ($\approx 4 - 9$ MeV) were too small to get over the Coulomb barrier that was known to exist at the nuclear surface. The Coulomb barrier is the electrostatic potential barrier which surrounds the nucleus and opposes both the entry and escape of positively charged particles (see Figure 11). In Rutherford's alpha particle scattering experiments it was found that the highest energy natural alpha particles that existed in his time, when approaching nuclei as close as their energies would allow, showed that the Coulomb potential continued up to a distance of about 3×10^{-14} m. This Coulomb potential, which is assumed to exist outside the nucleus, and an assumed nuclear potential are shown in Figure 11.

D. Theory of Alpha Decay

According to Gamow, Gurney, and Condon, who analyzed alpha decay from a quantum mechanical viewpoint in 1928, the alpha particle is assumed to exist as a separate entity within the nucleus prior to emission, confined by the nuclear potential shown in Figure 11. In the figure the vertical axis represents energy; the horizontal axis represents distance as measured from the center of the nucleus ($r = 0$). The potential function

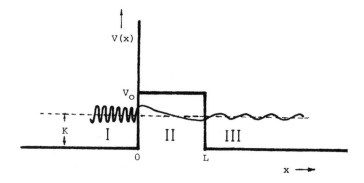

FIGURE 12. Particle incident upon a simple one-dimensional rectangular potential barrier.

$$V(r) = \frac{2Ze^2}{4\pi\epsilon_0 r}$$

represents an increasing repulsive force on the alpha particle as it approaches the nucleus. This is the alpha particle repulsion which was investigated by Rutherford. Inside the nucleus the exact shape of the potential energy curve is not known, but it must be attractive. For simplicity it is shown as a potential well. The kinetic energy of the alpha particle is K_a and the height of the potential barrier is V_B. (K_a ranges from 4 to 9 MeV; V_B is somewhat larger than this.) Since $V_B > K_a$, classical physics predicts that the alpha particle will never be able to get out of the nucleus but will bounce back and forth between the "walls" of the potential well forever. According to quantum mechanics, however, the alpha particle has a finite probability of escaping from the nucleus by *tunneling* through the potential barrier. For larger values of alpha particle energy, K_a, the thickness of the barrier, $R - R_0$, is smaller and the probability of escape therefore larger (short half-life); for smaller values of K_a, the thickness of the barrier is greater and the probability of escape is therefore smaller (long half-life). In the diagram, the alpha particle wave function is shown schematically with a large amplitude within the nucleus; it is greatly attenuated by the potential barrier; and it is shown with a very small amplitude beyond the barrier.

The probability of penetrating the barrier (called the *transparency*) is defined as:

$$transparency = \frac{transmitted\ intensity}{incident\ intensity}$$

We previously saw that the radioactive decay constant was given as:

$$\frac{-dN}{dt} = \lambda N$$

Writing this in the form:

$$\lambda = \frac{\frac{-dN}{N}}{dt}$$

we can think of λ as the decay probability per unit time. From a semiclassical point of view we can express λ as:

$$\lambda = \nu P$$

where ν is the number of collisions per second that the alpha particle makes with the wall of the potential barrier, and P is the probability that the alpha particle will penetrate the potential barrier per collision (the transparency). Hence:

$$\lambda = \text{rate of hitting the barrier} \times \text{transparency}$$

If we assume that the alpha particle within the nucleus is moving with a velocity of v_{in}, then

$$\nu = \frac{v_{in}}{2R_0}$$

where $2R_0$ is the nuclear diameter as shown in Figure 11.

In order to demonstrate the magnitude of the various quantities introduced above let us examine the alpha decay of $^{238}_{92}U$ which has a half-life of 4.51×10^9 years. This isotope of uranium emits alpha particles with energies of 4.19 MeV in the decay:

$$^{238}_{92}U \rightarrow ^{234}_{90}Th + ^{4}_{2}He$$

Now, assuming that the 4.19 MeV alpha particle exists in the $^{234}_{90}Th$ nucleus, it will have a velocity $v_{in} \simeq 1.42 \times 10^7$ m/sec. For the $^{234}_{90}Th$ nucleus, $R_0 \simeq 8.6 \times 10^{-15}$ m. Therefore:

$$\nu = \frac{v_{in}}{2R_0} = \frac{1.42 \times 10^7}{2 \times 8.6 \times 10^{-15}} \simeq 10^{21} \text{ collisions/sec}$$

It is astounding to realize that the alpha particle in this illustration collides with the potential barrier about 10^{21} times per second and still takes 4.5×10^9 years on the average to escape from the nucleus.

In order to understand the quantum mechanical phenomenon of barrier penetration upon which the theory of alpha particle emission is based, let us digress momentarily in order to examine the simple one-dimensional rectangular potential barrier problem from a quantum mechanical point of view.

Assume that a particle of rest mass m and kinetic energy K is incident upon a potential barrier of height V_0, and width L with $V_0 > K$ as shown in Figure 12.

As shown in the diagram, the one-dimensional potential can be described as:

$$V(x) = \begin{cases} 0 & x < 0 \\ V_0 & 0 < x < L \\ 0 & x > L \end{cases}$$

We assume that the particle is incident upon the potential barrier from the left. Classical mechanics predicts that in this problem with $K < V_0$ the particle has zero probability of moving from region I to region III. By calculating the transparency we will show that even with $K < V_0$ quantum mechanics allows the particle to penetrate the potential barrier:

$$\text{transparency} = \frac{\text{transmitted intensity}}{\text{incident intensity}} = \frac{\text{intensity in region III}}{\text{intensity in region I}}$$

To do this we will apply the one-dimensional, time-independent Schrödinger equation to each of the three regions. In region I:

$$\frac{d^2\psi_I}{dx^2} + \frac{2mK}{\hbar^2}\,\psi_I = 0$$

Let

$$\frac{\sqrt{2mK}}{\hbar} = k_I$$

Then the solution of the equation

$$\frac{d^2\psi_I}{dx^2} + k_I^2\,\psi_I = 0$$

can be written simply as:

$$\psi_I = Ie^{ik_I x} + Re^{-ik_I x}$$

where I represents the amplitude of the wave incident upon the potential and R represents the amplitude of the wave reflected from the potential at x = 0. In region II:

$$\frac{d^2\psi_{II}}{dx^2} - \frac{2m}{\hbar^2}\,(V_0 - K)\,\psi_{II} = 0$$

Let

$$\frac{\sqrt{2m(V_0 - K)}}{\hbar} = k_{II}$$

Then the solution of the equation

$$\frac{d^2\psi_{II}}{dx^2} - k_{II}^2\,\psi_{II} = 0$$

can be written simply as:

$$\psi_{II} = Ae^{k_{II} x} + Be^{-k_{II} x}$$

where A and B are constants.
In region III:

$$\frac{d^2\psi_{III}}{dx^2} + \frac{2mK}{\hbar^2}\,\psi_{III} = 0$$

Let

$$\frac{\sqrt{2mK}}{\hbar} = k_{III} = k_I$$

Then the solution of the equation

$$\frac{d^2\psi_{III}}{dx^2} + k_I^2\,\psi_{III} = 0$$

can be written simply as: $\psi_{III} = Te^{ik_I x}$ where T represents the transmitted amplitude. In the above solution we have dropped a term of the form $e^{-ik_I x}$ since that would represent a wave moving to the left in region III and we are assuming that only a transmitted wave exists there — a reflected wave or a source of particles traveling to the left does not exist.

In order to evaluate the constants I, R, A, B, and T we apply boundary conditions on ψ_I, ψ_{II}, and ψ_{III} which require that the eigenfunctions and their first derivatives must be continuous at $x = 0$ and $x = L$. These conditions may be stated as follows:

$$\psi_I(0) = \psi_{II}(0)$$

$$\frac{d\psi_I(0)}{dx} = \frac{d\psi_{II}(0)}{dx}$$

$$\psi_{II}(L) = \psi_{III}(L)$$

$$\frac{d\psi_{II}(L)}{dx} = \frac{d\psi_{III}(L)}{dx}$$

The application of the above boundary conditions leads to the following equations involving the arbitrary constants:

- $I + R = A + B$
- $ik_I(I - R) = k_{II}(A - B)$
- $Ae^{k_{II}L} + Be^{-k_{II}L} = Te^{-ik_I L}$
- $k_{II}(Ae^{k_{II}L} - Be^{-k_{II}L}) = ik_I Te^{ik_I L}$

Using the above equations we can eliminate A, B, and R and express I in terms of T:

$$I = \frac{T}{4} e^{ik_I L} \left[\left(1 + \frac{ik_I}{k_{II}}\right) \left(1 + \frac{k_{II}}{ik_I}\right) e^{-k_{II}L} + \left(1 - \frac{ik_I}{k_{II}}\right) \left(1 - \frac{k_{II}}{ik_I}\right) e^{k_{II}L} \right]$$

From this expression the transparency for the one-dimensional square well is found to be:

$$\text{Transparency} = \frac{TT^*}{I\,I^*} = \frac{|T|^2}{|I|^2}$$

$$= \frac{1}{\left[1 + \dfrac{V_0^2\,(e^{k_{II}L} - e^{-k_{II}L})^2}{16K\,(V_0 - K)} \right]}$$

$$= \frac{1}{\left[1 + \dfrac{V_0^2 \sinh^2 k_{II}L}{4K\,(V_0 - K)} \right]}$$

If $k_{II}L \gg 1$, then:

$$\text{Transparency} \approx \frac{16K}{V_0^2}\,(V_0 - K)\, e^{-2k_{II}L}$$

The coefficient multiplying the exponential term ranges in value between 0 (K = 0 and

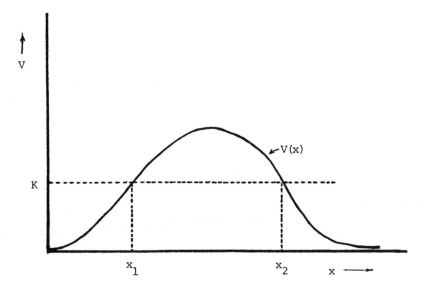

FIGURE 13.   A one-dimensional potential barrier of arbitrary shape showing the classical turning points x_1 and x_2.

$K = V_0$) and 4 ($K = V_0/2$). In most cases (for K not near zero and K not near V_0) the coefficient is usually taken as unity. With this approximation, the probability of tunneling through the potential barrier is written as: $P \simeq e^{-\gamma}$ where, in this case,

$$\gamma = 2k_{II}L = \frac{2\sqrt{2m(V_0 - K)}\, L}{\hbar}$$

The problem above was for a square potential barrier. For a barrier of arbitrary shape γ is written in terms of an integral:

$$\gamma = \frac{2}{\hbar} \int_{x_1}^{x_2} \sqrt{2m\{V(x) - K\}}\; dx$$

where x_1 and x_2 are the points at which $K = V(x)$ as shown in Figure 13.

We can now apply the results of the previous development to alpha decay:

Transparency $= P = e^{-\gamma}$ where

$$\gamma = \frac{2}{\hbar} \int_{R_0}^{R} \sqrt{2m\{V(r) - K\}}\; dr$$

where

$$V(r) = \frac{2Ze^2}{4\pi\epsilon_0 r}$$

as shown in Figure 11.

This integral can be solved, giving for the alpha decay problem:

$$\gamma = \frac{8Ze^2}{4\pi\epsilon_0 \hbar v} \left[\cos^{-1}\left(\frac{R_0}{R}\right)^{1/2} - \left(\frac{R_0}{R}\right)^{1/2}\left(1 - \frac{R_0}{R}\right)^{1/2} \right]$$

where v is the velocity of the emitted alpha particle and R and R_0 are as shown in Figure 11.

This expression can be written in terms of energy using the fact that at r = R,

$$K_\alpha = \frac{2Ze^2}{4\pi\epsilon_0 R}$$

Then

$$\frac{R_0}{R} = \frac{R_0}{\left(\frac{2Ze^2}{4\pi\epsilon_0 K_\alpha}\right)} = \frac{K_\alpha}{\left(\frac{2Ze^2}{4\pi\epsilon_0 R_0}\right)} = \frac{K_\alpha}{V_B}$$

where, as shown in Figure 11, V_B is the height of the potential barrier. Then:

$$\gamma = \frac{8Ze^2}{4\pi\epsilon_0 \hbar v} \left[\cos^{-1}\left(\frac{K_\alpha}{V_B}\right)^{1/2} - \left(\frac{K_\alpha}{V_B}\right)^{1/2}\left(1 - \frac{K_\alpha}{V_B}\right)^{1/2} \right]$$

We wrote previously that $\lambda = \nu P$ so that $\lambda = \nu e^{-\gamma}$ with γ given above. This result provides excellent agreement with observed disintegration rates (for even-even nuclei, and ground state to ground state transitions) and, as mentioned previously, was one of the first triumphs that helped to establish the new quantum mechanics in the late 1920s.

When the potential barrier is very high with respect to the kinetic energy of the alpha particle, i.e., $V_B \gg K_a$, we have the approximation:

$$\gamma \simeq \frac{8Ze^2}{4\pi\epsilon_0 \hbar v}\left(\frac{\pi}{2}\right)$$

and we can write:

$$\lambda = \nu e^{-\left(\frac{8Ze^2}{4\pi\epsilon_0 \hbar v}\right)\left(\frac{\pi}{2}\right)} = \nu e^{-(\text{constant})/v}$$

Using logarithms we can write the log λ in the form:

$$\log\lambda = A - B/v$$

where A and B are positive constants. Since

$$\lambda = \frac{\ln 2}{T_{1/2}}$$

$$\log\lambda = \text{constant} - \log T_{1/2}$$

Transferring from velocity to energy we can write:

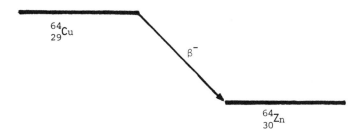

FIGURE 14.   Decay scheme for the β^- decay of $^{64}_{29}$Cu.

$$\frac{B}{v} = \frac{B'}{\sqrt{K_\alpha}}$$

so that

$$\log T_{1/2} = A' + \frac{B'}{\sqrt{K_\alpha}}$$

This expression, which relates the half-life of an alpha-unstable nucleus to the energy of the emitted alpha particle is in good agreement with experiment. It is in a form that provides a theoretical explanation for the very early observation (noted above) that the most energetic alpha particles are emitted from nuclei with the shortest half-lives and that the least energetic alpha particles are emitted from nuclei with the longest half-lives.

VI. A DETAILED ANALYSIS OF BETA DECAY

A. Beta Particle Spectra

The decay scheme for the β^- decay of $^{64}_{29}$Cu is shown in Figure 14.

The transition $^{64}_{29}$Cu $\rightarrow$ $^{64}_{30}$Zn + e$^-$ + $\bar{\nu}$ has a Q value of 0.573 MeV. If the β^- particles emitted during this decay were analyzed, the energy spectrum observed would be that shown in Figure 15.

Unlike the monoenergetic character of alpha particle spectra, we see that the beta particle spectrum is continuous. This is typical of all beta particle spectra. We see that a few beta particles possess a kinetic energy equal to the Q value (0.573 MeV), but that most of them do not. This is explained by the fact that beta decay is a three body process, i.e., three particles share the transition energy:

1.   The daughter nucleus
2.   The emitted beta particle
3.   The emitted antineutrino

As in the case of alpha decay, the daughter nucleus takes a negligible amount of the transition energy, almost all of it being shared between the beta particle and the antineutrino. When the antineutrino takes its maximum energy, the beta particle takes none; when the antineutrino takes no energy, the beta particle takes the maximum energy, labeled K_β(max) in Figure 15. In the general case then:

$$K_{\beta^-} + K_\nu = K_{\beta^-}(\text{max}) = K_{\bar{\nu}}(\text{max})$$

Historically, it was the continuous nature of beta particle spectra that led Pauli in

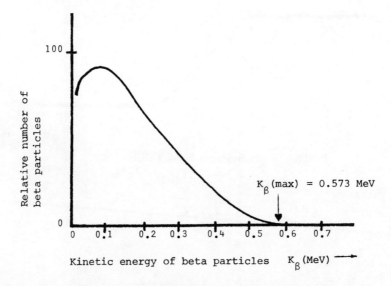

FIGURE 15.   Beta-particle spectrum for the decay of $^{64}_{29}$Cu.

1931 to postulate the existence of the neutrino a quarter of a century before it was actually observed. It was in this way that Pauli was able to preserve some very basic conservation laws in beta decay.

The energy K_β(max) — often referred to as the *end point energy* — is very important in beta decay, for it is very near to the disintegration energy, Q. (The difference between the two is the negligible recoil energy of the daughter nucleus.)

B. Beta Decay Energetics

In our generalized notation β^- decay, β^+ decay, and electron capture were written as:

$$\beta^- \text{ decay:} \qquad ^A_Z X \;\rightarrow\; ^{\;\;A}_{Z+1} Y \;+\; e^- \;+\; \bar\nu_e$$

$$\beta^+ \text{ decay:} \qquad ^A_Z X \;\rightarrow\; ^{\;\;A}_{Z-1} Y \;+\; e^+ \;+\; \nu_e$$

$$\text{Electron capture:} \quad ^A_Z X \;+\; e^- \;\rightarrow\; ^{\;\;A}_{Z-1} Y \;+\; \nu_e$$

Assuming that the parent nucleus in each case is at rest initially and that the neutrino has zero rest mass (it actually has a small rest mass) we can then write energy equations based on the above three transitions:

$$
\begin{aligned}
\beta^- \text{ decay:} \quad & M_X c^2 = M_Y c^2 + m_e c^2 + K_Y + K_{e^-} + K_\nu \\
& = M_Y c^2 + m_e c^2 + K_Y + K_{e^-}(\text{max}) \\
\beta^+ \text{ decay:} \quad & M_X c^2 = M_Y c^2 + m_e c^2 + K_Y + K_{e^+} + K_\nu \\
& = M_Y c^2 + m_e c^2 + K_Y + K_{e^+}(\text{max}) \\
\text{Electron capture:} \quad & M_X c^2 + m_e c^2 = M_Y c^2 + K_\nu
\end{aligned}
$$

The Q value in β^- decay is given as:

$$Q = K_Y + K_{e^-}(\text{max}) = M_X c^2 - M_Y c^2 - m_e c^2$$

Now, the above masses M_X and M_Y are nuclear masses. If we use atomic masses $M_X{}'$ and $M_Y{}'$ for convenience:

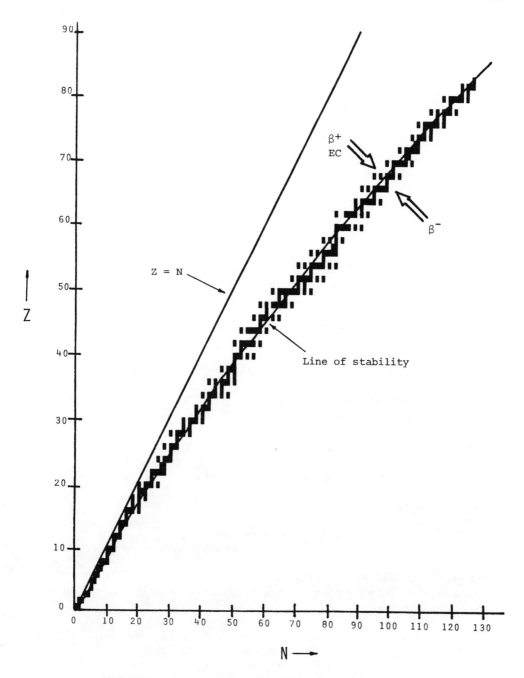

FIGURE 16. Graph of beta-stable nuclei showing the line of stability.

$$M_x' = M_x + Zm_e - B_x/c^2$$

and

$$M_Y' = M_Y + (Z + 1)m_e - B_Y/c^2$$

where B_x and B_Y are the atomic electron binding energies of the parent and daughter atoms, respectively.

Then:

$$Q = K_\gamma + K_{e^-}(\text{max})$$
$$= \{(M_x' - Zm_e + B_x/c^2) - [M_y' - (Z + 1)m_e + B_y/c^2] - m_e\}c^2$$
$$= \{M_x' - M_y'\}c^2$$

where we have ignored the small difference in atomic electron binding energy between the parent and daughter atoms.

For spontaneous β^- decay Q must be positive, so we see that the condition for spontaneous β^- decay is that $M_x' > M_y'$, i.e., the atomic mass of the parent nuclide must exceed that of the daughter.

The Q value in β^+ decay is given as:

$$Q = K_\gamma + K_{e^+}(\text{max}) = M_x c^2 - M_y c^2 - m_e c^2$$

Again, using atomic masses:

$$M_x' = M_x + Zm_e - B_x/c^2$$
$$M_y' = M_y + (Z - 1)m_e - B_y/c^2$$

Then:

$$Q = K_\gamma + K_{e^+}(\text{max})$$
$$= \{(M_x' - Zm_e + B_x/c^2) - [M_y' - (Z - 1)m_e + B_y/c^2] - m_e\}c^2$$
$$= \{M_x' - M_y' - 2m_e\}c^2$$

where the small difference between the atomic electron binding energy of the parent and daughter atoms has been ignored.

The condition for spontaneous β^+ decay, then, is: $M_x' > (M_y' + 2m_e)$, i.e., the atomic mass of the parent nuclide must exceed that of the daughter by at least two electron masses.

The Q value in electron capture is given as:

$$Q = K_\nu = M_x c^2 + m_e c^2 - M_y c^2$$

Using atomic masses:

$$M_x' = M_x + Zm_e - B_x/c^2$$
$$M_y' = M_y + (Z - 1)m_e - B_y/c^2$$

However, the daughter atom is produced with a vacancy in one of its electron shells and so is in an excited state which exceeds the ground state by E_B, the binding energy of the captured electron:

$$Q = \{(M_x' - Zm_e + B_x/c^2) + m_e - [M_y' + E_B/c^2 - (Z - 1)m_e + B_y/c^2]\}c^2$$
$$= (M_x' - M_y')c^2 - E_B$$

The condition for electron capture, then, is that: $M_x' > M_y' + E_B/c^2$, i.e., the atomic mass of the parent nuclide must exceed that of the daughter plus the binding energy of the captured electron.

Remember that electron capture and positron emission are competing processes — however, they do have different conditions for their occurrence. Consequently, there are many transitions which occur via electron capture for which positron emission is energetically forbidden.

As an example, let us examine the decay of $_4^7$Be to $_3^7$Li. The atomic mass of $_4^7$Be = 7.016930u. The atomic mass of $_3^7$Li = 7.016004u. Therefore, Δm = 7.016930 −

7.016004 = 0.000926u and we see that only electron capture is allowed in this transition since Δm is less than two electron masses ($2m_e$ = 0.001097u).

The systematics of beta decay can be pictured diagrammatically by referring to the chart of the nuclides as shown in Figure 16.

This figure shows the *line of stability,* the systematic location on the chart of the nuclides of the β-stable nuclei. We see that nuclei below the line of stability (excess neutrons) decay via β^- emission in order to reach stability; those above the line of stability (excess protons) decay via β^+ emission and electron capture in order to achieve stability.

We also see that only the lightest nuclei tend to have equal numbers of protons and neutrons. As the mass increases, stable nuclei tend to have more neutrons than protons. This is a consequence of the increased coulomb repulsion as Z increases, which tends to reduce the binding energy making the nucleus unstable. The effect is countered by having more neutrons than protons which produces additional attractive forces that tend to increase the binding energy.

C. Fermi Theory of Beta Decay

The first successful theory of beta decay was developed by Fermi in 1934 incorporating the neutrino hypothesis of Pauli. Fermi assumed that the basic transformations were

$$n \rightarrow p + \beta^- + \bar{\nu} \qquad (\beta^- \text{ decay})$$
$$p \rightarrow n + \beta^+ + \nu \qquad (\beta^+ \text{ decay})$$

and

$$p + e^- \rightarrow n + \nu \qquad (\text{electron capture})$$

The introduction of the neutrino into the theory provided an explanation of the continuous energy spectrum that was observed in beta decay, for the neutrino could carry away the energy that seemed to be missing in beta ray spectra. The above transitions suggest the following properties for the neutrino:

1. Zero charge. (Only in this way will there be conservation of charge in each transformation.)
2. An intrinsic spin of ½ ($h/2\pi$) where h is Planck's constant. (Only in this way will there be conservation of angular momentum in each transformation. The neutrino is thus assumed to be a fermion which follows Fermi-Dirac statistics — the statistics of spin one-half particles.)
3. Very small mass. (We have already seen that the conversion of known masses to energy accounts for the observed end-point energies within experimental error.)

Fermi assumed that the electron and neutrino did not preexist within the nucleus, but were created at the time of the disintegration by an interaction which exists between the nucleon, electron, and neutrino. This interaction — now called the *weak interaction* — has an extremely short range.

The Fermi theory yields the following expression for the probability per unit time that a beta particle with momentum between p and p + dp will be emitted:

$$N(p)dp = \text{constant} \times g^2 \; |M|^2 \times (K_{max} - K_\beta)^2 \; p_\beta^2 \; dp_\beta$$

where g is a universal constant which characterizes the weak interaction called the beta-decay coupling constant. It is of an order of magnitude of 10^{-62} J-m^3. K_{max} is the total energy available for the disintegration (equal, essentially, to the end-point energy). K_β and p_β are the kinetic energy and momentum, respectively, of the beta particle, and $|M|^2$ represents the absolute square of the quantum mechanical matrix element which describes the transformation. The matrix element is defined as

$$M = \int \psi_f^* \psi_i d\tau$$

where ψ_i represents the initial wave function of the parent nucleus and ψ_f^* represents the complex conjugate of the wave function of the daughter nucleus. Hence | M | depends only on nuclear properties. Simplifications made by Fermi allowed him to factor out the effects of the electron and neutrino. (This is the origin of the coupling constant g in the most simple case that Fermi considered.)

The previous expression for N(p)dp must be modified to account for the distortion of the wave function of the electron or positron by the Coulomb force of the nucleus. This distortion is especially pronounced for low energy beta particles where the emission of negative beta particles will be reduced and the emission of positive beta particles will be increased. Taking this effect into consideration:

$$N(p)dp = \text{constant} \times g^2 \times |M|^2 \, F(Z,K_\beta) \, (K_{max} - K_\beta)^2 \, p_\beta^2 \, dp_\beta$$

where the function $F(Z,K_\beta)$, called the *Fermi function,* takes into account the nuclear Coulomb effects. $F(Z,K_\beta)$ has the value unity for $Z = 0$ (no Coulomb interaction). Z is taken as negative for β^- emitters and positive for β^+ emitters to allow for the differing effects.

The above expression was found to be in excellent agreement with the shape of observed beta momentum spectra for a class of beta transitions called *allowed transitions.* An extremely sensitive test of the Fermi theory involves what is called a *Kurie-plot.* In utilizing this technique, the expression for N(p) dp is rewritten as:

$$\left[\frac{N(p)}{F(Z,K_\beta)p_\beta^2} \right]^{1/2} = \text{constant} \times g \times |M| \, (K_{max} - K_\beta)$$

Since, as mentioned above, |M| depends only on nuclear properties, it is independent of electron energy and can be considered a constant. Therefore:

$$\left[\frac{N(p)}{F(Z,K_\beta) \, p_\beta^2} \right]^{1/2} \propto (K_{max} - K_\beta)$$

and we see that the quantity

$$\left[\frac{N(p)}{F(Z,K_\beta) \, p_\beta^2} \right]^{1/2}$$

plotted against K_β should result in a straight line. Moreover, this straight line should intersect the energy axis at $K_\beta = K_{max}$. Hence, the Kurie plot turns out to be an excellent and sensitive method for determining end-point energies. The Kurie plot provides excellent confirmation of the Fermi theory of beta decay for allowed transitions. For so-called *forbidden transitions,* which are characterized by long half-lives, the above expression for N(p)dp does not predict a spectral shape that is in good agreement with observation. The Fermi theory, however, can be refined by reconsidering the simplifying assumptions made above (particularly with regard to the form of the matrix element, | M |) to attain good agreement between theory and experiment.

N(p) dp is the probability per unit time that a beta particle with momentum between p and p + dp will be emitted. The total probability per unit time that a beta particle will be emitted with any momentum is the disintegration constant, which is the integral over all possible values of p:

$$\lambda = \frac{\ln 2}{T_{1/2}} = \int_0^{p_{max}} N(p)dp$$

$$= \int_0^{p_{max}} \text{constant} \times g^2 \times |M|^2 \, F(Z,K_\beta) \, (K_{max} - K_\beta)^2 \, p_\beta^2 \, dp_\beta$$

$$= \text{constant} \times g^2 \times |M|^2 \int_0^{p_{max}} F(Z,K_\beta) \, (K_{max} - K_\beta)^2 \, p_\beta^2 \, dp_\beta$$

The integral above can be evaluated numerically and the result expressed as a function of Z and K_{max}:

$$\int_0^{p_{max}} F(Z,K_\beta) \, (K_{max} - K_\beta)^2 \, p_\beta^2 \, dp_\beta = f(Z,K_{max})$$

Then:

$$\lambda = \frac{\ln 2}{T_{1/2}} = \text{constant} \times g^2 \, |M|^2 \, f(Z,K_{max})$$

and:

$$f(Z,K_{max}) \, T_{1/2} = \frac{\text{constant}}{g^2 |M|^2}$$

The product $f(Z,K_{max})T_{1/2}$ — usually written as ft — is called the *comparative life time* or the *ft value* and can be evaluated for a particular beta decay transition by measuring the half-life and calculating $f(Z,K_{max})$. In cases where | M $|^2$ is approximately equal to unity the ft value can be used to estimate the magnitude of the coupling constant g. Assuming that the value of g is known, the expression ft = constant/| M $|^2$ can be used to determine information about the matrix element M by means of a half-life measurement. This can lead to the classification of a beta transition in terms of *degrees of forbiddenness*. Since ft values vary through such an enormous range — from about 10^3 to about 10^{18} sec — it is commonplace in beta spectroscopy to use the $\log_{10}$ ft. Thus decay schemes often show log ft values for beta transitions. When log ft values are analyzed for the known beta emitters, distinct groups are observed. The smallest values apply to a group of light nuclei for which log ft is about 2.7 to 3.7. This suggests a strong overlap in the wave functions describing the initial and final nuclear states in this group. This implies a large matrix element, M, which produces a small log ft value. Such transformations are described as *favored* or *superallowed transitions*. Another group of transitions is found for which values of log ft lie between about 4 and 6.1.

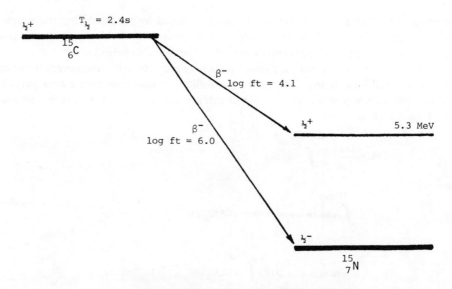

FIGURE 17.   Decay scheme showing the β^- decay of $^{15}_{6}C$.

These transformations are designated as *allowed transitions*. Another group of transitions is found for which the values of log ft range from about 6 to 9. These are designated as *first forbidden transitions*.

The simplified theory described above, though accounting for the shape of the beta spectrum in allowed transitions and showing the relation between energies and life times by means of the ft values, does not apply to forbidden transitions. However, it can be extended to include them. Moreover, although Fermi assumed the conservation of parity in his theory, the theory can be modified to include basic changes resulting from the discovery in 1956 of the nonconservation of parity in beta decay. Parity, a mathematical property of a wave function, refers to whether or not a wave function changes sign if the space coordinates change sign, i.e., if $\psi(x,y,z) = \psi(-x,-y,-z)$ the wave function is said to have even (+) parity; and if $\psi(x,y,z) = -\psi(-x,-y,-z)$ the wave function is said to have odd (−) parity.

In decay schemes, the parity of nuclear states is usually shown by a " + " or "−" sign.

As an example of parity conservation in beta decay consider the decay scheme shown in Figure 17 which describes the decay of $^{15}_{6}C$ to $^{15}_{7}N$. The parity and angular momentum assignments are shown. (Angular momenta are given in units of $\hbar$.) One beta transition is from the ½ + ground state of $^{15}_{6}C$ to the ½ + excited state of $^{15}_{7}N$. If parity were conserved, this transition would result in the emission of a beta particle with an angular momentum of even parity. (The intrinsic parity of the electron is even.) In the transition from the ½ + ground state of $^{15}_{6}C$ to the ½ − ground state of $^{15}_{7}N$ the parity of the nuclear states changes. For parity to be conserved in this transition the angular momentum of the emitted beta particle would have to have odd parity.

Until 1956 such a conservation of parity law was considered to hold in all processes. Since then, however, it has been discovered that the conservation of parity does not hold in processes involving the weak interaction, e.g., beta decay.

If the beta particle and the antineutrino are emitted from a nucleus with their spins in opposite directions (antiparallel), then — if they carry no orbital angular momentum — they will carry away no net angular momentum from the nuclear system. This situation leads to a *selection rule* by Fermi, for, in order to conserve angular momentum in such a beta decay the nuclear angular momentum cannot change. I.e.,

Table 1

SELECTION RULES IN BETA DECAY

Nature of transition	ΔI	Parity change	Selection rules	Log $_{10}$ ft
Superallowed	0	No	Fermi	2.7—3.7
Allowed	0	No	Fermi	4—6
	0,±1 (No 0→0)	No	Gamow-Teller	
First forbidden	0,±1 (No 0→0)	Yes	Fermi	6—10
	0,±1,±2 (No 0→0)	Yes	Gamow-Teller	

.
.
.

Etc.

$$\Delta I = 0 \text{ (Fermi selection rule)}$$

where ΔI represents the change in angular momentum from parent to daughter nucleus.

If the beta particle and the antineutrino — with no orbital angular momentum — are emitted from the nucleus with spins parallel, they carry off one unit of angular momentum from the nuclear system. This situation leads to a selection rule by Gamow and Teller who extended Fermi's theory to take spins of the emitted particles into account:

$$\Delta I = 0 \text{ or } \pm 1 \text{ (but not } 0_i \to 0_f)$$
$$\text{(Gamow-Teller selection rule)}$$

This results from the fact that the angular momentum is a vector quantity: $\Delta \vec{I} = \vec{I}_f - \vec{I}_i = 1$. (We see from this relationship why $I_f = 0$ and $I_i = 0$ cannot be allowed.)

In both the Fermi and Gamow-Teller cases above, the nuclear state must not change parity.

Beta decay can occur even when ΔI is larger than one because the beta particle and antineutrino can be emitted with orbital angular momentum, which, thus far, has not been considered. However, larger changes in angular momentum result in higher degrees of forbiddenness. This is shown in Table 1, which summarizes many of the previously developed concepts in beta decay.

D. Ratio of Electron Capture to Positron Emission

The previous theory can also be applied to electron capture, and with slight modifications the decay constant λ_{EC} can be determined. This corresponds to the beta decay disintegration constant from the above Fermi theory, λ, which we will now write as λ_β.

As we have seen, the decay constant λ_{β^+} involves unknown nuclear matrix elements. The same is true of the λ_{EC} decay constant. Therefore, we can not compare either of these expressions directly with experimental observations. However, it was mentioned previously that electron capture generally competes with positron decay in neutron deficient nuclei. Since the absolute square of the matrix element $| M |^2$ which describes each of these competing processes depends only upon the initial and the final nuclear states, the matrix elements are the same in each case. The ratio $\lambda_{EC}/\lambda_{\beta^+}$ of the probability for electron capture to positron emission is therefore independent of the nuclear matrix elements. This is important because it means that the ratio $\lambda_{EC}/\lambda_{\beta^+}$ can be observed experimentally. It was mentioned previously that it is the innermost or K-shell electrons that have the greatest probability of being captured. Therefore, it is the ratio $\lambda_K/\lambda_{\beta^+}$ that is most easily investigated experimentally. Reasonably good agreement has been found between theory and experiment which provides support for the theory of beta decay.

Experimental observations of the electron capture process differ from beta spectroscopy. Since the actual electron capture cannot be observed directly it is usually the subsequent X-ray or Auger electron that is observed experimentally. Moreover, if the daughter nucleus is left in an excited state following electron capture, a gamma ray may also be observed.

Experimental values for $\lambda_K/\lambda_{\beta^+}$ show that this ratio increases with Z and decreases with disintegration or end-point energy. Hence, for low-Z nuclides with relatively large disintegration energies, K-capture is less probable than positron emission. For high-Z nuclides K-capture is favored over positron emission, particularly for smaller end-point energies. Physically, these results can be explained by the fact that as Z increases, electron orbits become smaller so that there is an increasing probability of an atomic electron being inside the nucleus. Also, as Z increases the increasing potential barrier will tend to inhibit positron emission.

VII. A DETAILED ANALYSIS OF GAMMA DECAY

A. Gamma Ray Spectra

We have already seen that gamma ray emission (as well as internal conversion) occurs when a nucleus undergoes a transformation from a state of higher energy to a state of lower energy without changing its proton number or neutron number. We saw that gamma ray energies are discrete:

$$E_\gamma = \Delta E = E_i - E_f = h\nu$$

(where we have ignored the recoil of the nucleus).

B. The Theory of Gamma Ray Emission

As a preparation for the theory of gamma ray emission it is helpful to review some basic concepts from classical electricity and magnetism, particularly the classical idea of multipole moments. We will simplify the following development by focusing on the electric dipole.

If two electric charges $+q$ and $-q$ are separated by a distance d, they constitute what is called an *electric dipole*. See Figure 18.

A quantity p, called the *electric dipole moment* is defined for this configuration:

$$p = qd$$

where p is the product of the charge and the separation distance. The electric dipole moment is often written as a vector quantity $\vec{p} = q\vec{d}$ where the direction of the vector d is taken as pointing from the negative to the positive charge.

We can extend the definition of the electric dipole moment for an arbitrary distribution of point charges:

$$p_x = \sum_i q_i x_i$$

$$p_y = \sum_i q_i y_i$$

$$p_z = \sum_i q_i z_i$$

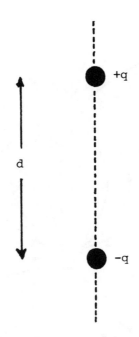

FIGURE 18.   Electric dipole with a dipole moment p = qd.

Similarly, we can express the *quadrupole moment* of an arbitrary distribution of point charges as:

$$p_{xx} = \sum_i q_i x_i^2$$

$$p_{xy} = \sum_i q_i x_i y_i \qquad \text{etc.}$$

which is a tensor quantity with six components. We can define the *octupole moment* of an arbitrary distribution of point charges as:

$$p_{xxx} = \sum_i q_i x_i^3$$

$$p_{xxy} = \sum_i q_i x_i^2 y_i \qquad \text{etc.}$$

which would be made up of nine components.

Note that the dimensionality of the various *multipoles* is x^l where $l = 0, 1, 2, 3, ...$ is called the *order* of the multipole. For $l = 0$ we would have a *monopole* which would be represented by a single point charge at the origin of coordinates.

For a continuous distribution of charges we would have

$$p_x = \int \varrho x \, dv; \; p_y = \int \varrho y \, dv; \; p_z = \int \varrho z \, dv,$$

where the charge densities, ϱ, may be functions of the coordinates.

Similarly for quadrupole moments, etc.

The familiar electric field surrounding the electric dipole is shown diagrammatically in Figure 19.

We can write the potential of an arbitrary charge distribution for any point external to the distribution. It is customary to expand this potential by means of a three-dimen-

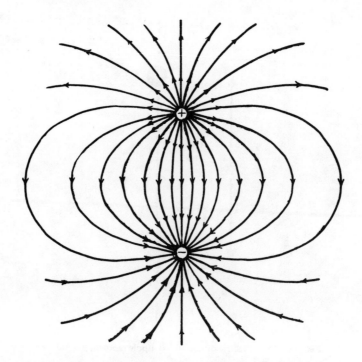

FIGURE 19.   Static electric field surrounding an electric dipole.

sional Taylor series. The resulting expansion is called the *multipole expansion* for it expresses the potential in terms of the successive multipole moments, i.e., monopoles, dipoles, quadrupoles, etc. This technique is important for it allows us to visualize *any* arbitrary distribution of electric charge as a collection of monopoles, dipoles, quadrupoles, etc. The various multipoles are often referred to as 2^t poles such that:

- 2^0-pole corresponds to a monopole
- 2^1-pole corresponds to a dipole
- 2^2-pole corresponds to a quadrupole
- 2^3-pole corresponds to an octupole
- 2^4-pole corresponds to a sixteen pole
- Et cetera

In a similar way we can consider static *magnetic multipoles* beginning with the magnetic dipole. This is most simply visualized as a steady-state current loop:

This configuration is referred to as a magnetic dipole because the magnetic field associated with it is analogous to the electric field which surrounds the electric dipole. This can be seen in Figure 20.

In a similar development to that above the magnetic potential of an arbitrary source of magnetism can be expanded in a Taylor series to produce a magnetic multipole expansion which expresses the magnetic potential in terms of the various magnetic

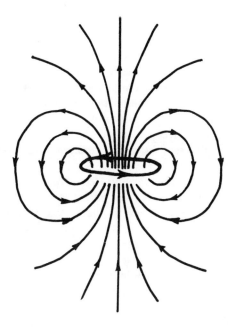

FIGURE 20. Static magnetic field associated
with a steady-state current loop.

multipole moments, i.e., magnetic dipoles, magnetic quadrupoles, magnetic octupoles,
etc. (In the magnetic multipole expansion there is no magnetic monopole since isolated
magnetic poles do not exist.)

The various magnetic multipoles are also referred to as 2^l-poles. When referring to
both electric and magnetic 2^l poles, the following convention is often used:

- E0 — electric monopole
- E1 — electric dipole
- M1 — magnetic dipole
- E2 — electric quadrupole
- M2 — magnetic quadrupole
- E3 — electric octupole
- M3 — magnetic octupole
- E4 — electric sixteen pole
- M4 — magnetic sixteen pole
- Et cetera

In summary, then, any arbitrary distribution of electric charge can be represented
in terms of the various electric multipoles; and any arbitrary distribution of steady-
state current loops can be represented in terms of the various magnetic multipoles. It
should be emphasized that for the arbitrary distribution and its equivalent multipole
representation, the external electric and magnetic fields are identical.

Thus far we have dealt with static charge distributions and steady-state current
loops. The external fields from such configurations will be static. As we have indicated
before, however, Maxwell, in his theory of electromagnetism, indicated that accelerated charges produce radiation. Accordingly, from a classical standpoint we can see
that as the electric charges — or the currents in the current loops — in the distributions
considered above begin to oscillate, radiation will be emitted. Radiation will also be
produced if current loops rotate or oscillate with respect to one another. All these cases

involve the acceleration of charged particles and classical electrodynamics would predict the production of radiated electromagnetic waves — that is, time-varying electric and magnetic fields will be set up at distant points in space.

Now, a most important point is that each radiating multipole at the source emits a *characteristic* radiation field. Consequently, radiation emitted from a variety of multipoles can be grouped according to the particular electric or magnetic multipole that emitted it. We refer, then, to *electric multipole radiation* and *magnetic multipole radiation* of a particular order. (Keep in mind that the fields from a particular *static* multipole are completely different from the *radiated* fields that are emitted when the multipole is time-varying.)

One of the properties of the time-varying fields from a radiating multipole that, in part, makes the fields characteristic is their parity. As discussed previously, parity refers to symmetry or reflective properties. In the present case it refers to the symmetry of the radiated fields.

For multipole radiation it is found that:

1.   The parity of electric multipole radiation $(E\ell) = (-1)^\ell$
2.   The parity of magnetic multipole radiation $(M\ell) = -(-1)^\ell$

where ℓ is the order of the multipole (2^ℓ).
Hence:

- E1 radiation has odd parity $(-)$
- E2 radiation has even parity $(+)$
- M1 radiation has even parity $(+)$
- M2 radiation has odd parity $(-)$
- Et cetera

In order to apply the previous ideas to the nucleus the classical expressions developed above must be written in terms of quantum mechanics. The x-component of the electric dipole moment for a continuous charge distribution which was written classically as

$$p_x = \int \varrho x \, dv$$

is written in quantum mechanics as an expectation value $<p_x>$ which is proportional to

$$e\int \psi_f^* x\psi_i dv$$

where ψ_f refers to the final state and ψ_i to the initial state. (These are time-dependent total wave functions.) When a nucleus is emitting radiation, it is under the influence of a time-dependent potential. Therefore the nuclear system changes in time from ψ_i to ψ_f. This change is governed by an *electric dipole moment transition matrix element* which is proportional to $\int \psi_f^* x\psi_i dv$. (These are time-independent eigenfunctions here.)

In order to establish the importance of the parity selection rule let us analyze the symmetry properties of this matrix element in detail. The integrand is either an even function or an odd function. If the integrand is an odd function (overall odd parity) then the integral will vanish, for a contribution to the integral at the coordinates x, y, and z will, because of the odd parity of the integrand, be exactly negated by the contribution to the integral at the coordinates −x, −y, and −z. Hence the integrand will be nonvanishing only if it is an even function (overall even parity). Now, the parity of x is odd since x = −(−x), so we see that for the integral to be nonvanishing the parities of the initial and final states must differ. This is the origin of a *parity selection rule*. We can also analyze the above integral on the basis of the previously mentioned parity of E1 radiation, which was $(-1)^1 = -1$ or odd. This means that the parity of the states must change in order that E1 radiation be emitted. (If the matrix element above had been the magnetic dipole moment transition matrix element then, for the integral to be nonvanishing, the initial and final nuclear states must have the same parity.)

The parity selection rule outlined above is exceedingly important for two reasons. First, although we saw that parity was not conserved in beta decay, it should be emphasized that it is strictly conserved in gamma decay (and, in fact, in all processes involving the so-called coulomb interaction). The parity selection rule can be written as

$$\Pi_i = \Pi_f \Pi_\gamma$$

where Π_i and Π_f represent the parity of the initial and final state, respectively, and Π_γ represents the parity of the emitted gamma radiation. The above expression is a statement of the conservation of parity.

The second reason why the parity selection rule is important is that the multipole moment transition matrix elements are related to the gamma ray decay constant (or transition probability). The gamma ray decay constant $\lambda (= \ln2/T_{1/2})$ for electric dipole radiation is proportional to

$$\{|\int \psi_f^* x\psi_i dv|^2 + |\int \psi_f^* y\psi_i dv|^2 + |\int \psi_f^* z\psi_i dv|^2\}$$

where each term in the above expression represents the square of a transition matrix element. Now if the parity of ψ_i and ψ_f were the same, each term in the above expression would vanish, signifying that a decay from state ψ_i to state ψ_f would be impossible via electric dipole radiation (i.e., $T_{1/2} = \infty$).

There is another selection rule that is essential to the understanding of gamma ray emission. This rule stems from the fact that each multipole moment of order ℓ produces photons with angular momentum ℓh. The law of conservation of angular momentum dictates that the total angular momentum of the initial system be equal to that of the final system (nucleus plus radiation). This leads to the selection rule on angular momentum:

$$\vec{I_i} = \vec{I_f} + \vec{\ell}$$

where $\vec{I_i}$ and $\vec{I_f}$ represent the angular momenta of the initial and final states, respectively, and $\vec{\ell}$ the angular momentum of the emitted photon. This is vector equation, and is sometimes written in the form

$$\vec{I_i} - \vec{I_f} = \vec{\ell}$$

The maximum value of the left-hand side is $|\vec{I_i}| + |\vec{I_f}|$; the minimum value is $|\vec{I_i}| - |\vec{I_f}|$ (or $|\vec{I_f}| - |\vec{I_i}|$ depending upon whether $|\vec{I_i}| > |\vec{I_f}|$ or $|\vec{I_f}| > |\vec{I_i}|$). Therefore, we can use the scalars $I_f (= |\vec{I_f}|)$ and $I_i (= |\vec{I_i}|)$ to write the angular momentum selection rule in another form: multipole radiation of order ℓ can be emitted in a transition between two nuclear states ψ_i and ψ_f only if

$$|I_i - I_f| \leqslant \ell \leqslant I_i + I_f$$

Since there is no multipole radiation with $\ell = 0$, the above selection rule absolutely forbids radiative transitions between two states for which $I_i = I_f = 0$.

We can summarize the above selection rules:

Table 2
PROPERTIES OF THE VARIOUS MULTIPOLE ORDERS OF GAMMA RADIATION

Type of radiation	Symbol	Angular momentum of emitted photon	Change in parity	$T_{1/2}$ for 1 MeV gamma ray
Electric dipole	E1	$1\hbar$	Yes	3×10^{-16} sec
Magnetic dipole	M1	$1\hbar$	No	2×10^{-14} sec
Electric quadrupole	E2	$2\hbar$	No	1×10^{-11} sec
Magnetic quadrupole	M2	$2\hbar$	Yes	1×10^{-9} sec
Electric octupole	E3	$3\hbar$	Yes	1×10^{-4} sec
Magnetic octupole	M3	$3\hbar$	No	1×10^{-2} sec

FOR ELECTRIC MULTIPOLE RADIATION
$|I_i - I_f| \leqslant \ell \leqslant I_i + I_f$ (not $I_i = I_f = 0$)
The parity of the nuclear states must change if ℓ is odd.
The parity of the nuclear states must not change if ℓ is even.

FOR MAGNETIC MULTIPOLE RADIATION
$|I_i - I_f| \leqslant \ell \leqslant I_i + I_f$ (not $I_i = I_f = 0$)
The nuclear parity must change if ℓ is even.
The nuclear parity must not change if ℓ is odd.

We saw above that when a multipole moment transition matrix element vanishes it signifies that that mode of decay is completely forbidden. Although one mode of multipole radiation may be forbidden, the next higher multipole radiation may be allowed. However, the probability of multipole emission decreases rapidly with increasing ℓ. Also, for a given value of ℓ, the emission of electric multipole radiation is much more probable than the emission of magnetic multipole radiation.

The above ideas are summarized in Table 2, which gives the angular momentum carried off by a particular multipole radiation, the parity change of the nuclear states in order that a particular multipole radiation might be emitted, and an estimate of the half-life of the transition when a 1 MeV photon is emitted (the half-life is extremely energy dependent, increasing rapidly as the energy decreases).

The marked increase in half-life in the last column of the table indicates the decrease in probability of higher order multipole emissions. As a result, in a given transition, one, and at most two, multipole radiations are of significance; very often only photons emitted with the lowest allowed angular momentum are of importance.

In the case of atomic transitions, only electric dipole radiation is important. It is possible for atoms to deexcite by emitting higher order multipole radiation, but this rarely happens since it is much more probable that the atom will deexcite through a collision. For the nucleus, this is not the case. First, the probability of higher multipole radiation does not fall off as rapidly as it does for atoms. Secondly, Coulomb repulsion prevents nuclei from getting too close to one another. Therefore, if selection rules prevent a nucleus from emitting lower order multipole radiation, the nucleus is usually forced to wait until it can decay via higher order multipole radiation with its longer half-life (unless it is able to release its excess energy by the process of internal conversion).

If the half-life of a nuclear state is long enough to be measured directly (which can happen for lower energy gamma rays for which low order multipole radiation is forbidden) the nuclear state is called an *isomeric state* or an *isomer*. (Sometimes the term *metastable state* is used.) Some isomeric states have lifetimes of the order of years.

75

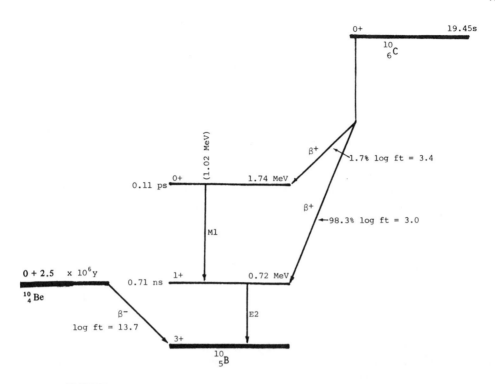

FIGURE 21.   Decay scheme showing the decay of $^{10}_{4}$Be and $^{10}_{6}$C to the stable nuclide $^{10}_{5}$B.

On an energy level diagram an isomeric state is denoted by the superscript m following the mass number: $^{Am}_{Z}$ X.

Example — ^{80m}Br represents a long-lived (4.5 hr) isomeric state of ^{80}Br.

On an energy level diagram an isomeric transition is denoted by the letters IT.

As an illustration of some of the ideas we have covered, let us examine the decay scheme of $^{10}_{4}$Be and $^{10}_{6}$C to the stable nuclide $^{10}_{5}$B. This decay scheme is shown in Figure 21.

1.   For the β^- decay of $^{10}_{4}$Be note the high degree of forbiddenness as a result of the 0 + initial state and 3 + final state. The high degree of forbiddenness and the resulting long half-life for the decay of $^{10}_{4}$Be (2.5 × 10⁶ years) are suggested by the large log ft value of 13.7.

2.   For the β^+ decay to the 0 + excited state and the 1 + excited state of $^{10}_{5}$B we have allowed transitions which are indicated by the 3.4 and 3.0 values for log ft. The short lifetime of 19.45 sec is consistent with this. Note the absence of a transition from $^{10}_{6}$C to the ground state of $^{10}_{5}$B. The 0 + → 3 + change in angular momentum makes this transition highly forbidden.

3.   So, the 0 + and 1 + states of $^{10}_{5}$B are populated by the β^+ decay of $^{10}_{6}$C. As the $^{10}_{5}$B nucleus deexcites there are two gamma rays emitted, a 1.02 MeV gamma ray in the 0 + → 1 + transition and a 0.72 MeV gamma ray in the 1 + → 3 + transition. Note the absence of a gamma ray from the 0 + state directly to the 3 + ground state. The large ΔI decreases the probability of this transition. From our angular momentum selection rule the *lowest* order multipole that could bring about this transition would be $l = 3$. From consideration of the parity selection rule the radiation must have even parity; hence, the first multipole radiation that could satisfy the selection rules for this transition would be M3, or magnetic octupole radiation, so there is a very low probability that this transition will oc-

cur. The 1.02 MeV gamma ray is from a radiating M1 multipole — the parity given by $-(-1)^1 = +$ and the angular momentum change of $\Delta I = 1$ are consistent with the $0+ \rightarrow 1+$ transition. The relatively large energy of the gamma ray and the $\Delta I = 1$ change result in a lifetime of 0.11 psec for this state. The 0.72 MeV gamma ray to the ground state is from a radiating E2 multipole (E1 is disallowed by both the parity selection rule and the angular momentum selection rule). The parity for E2 radiation, $(-1)^2 = +$, and the angular momentum change of $\Delta I = 2$ are consistent with the $1+ \rightarrow 3+$ transition.

$I_i = 0 \rightarrow I_f = 0$ TRANSITIONS — It was pointed out previously that $I_i = 0 \rightarrow I_f = 0$ transitions were strictly forbidden in gamma decay. This is true since there is no multipole radiation of order $l = 0$. $I_i = 0 \rightarrow I_f = 0$ transitions can occur, however, by the process of internal conversion. Such transitions provide convincing evidence that internal conversion is, indeed, a one-step process.

C. Internal Conversion Coefficients

In a previous discussion it was pointed out that internal conversion was a decay mode which competed with gamma ray emission. This is true except for the $I_i = 0 \rightarrow I_f = 0$ transitions mentioned in the preceding section for which gamma decay is strictly forbidden.

Let us consider an excited nucleus whose only possible decay modes are gamma decay and internal conversion. The total decay constant for deexcitation of the state is given by:

$$\lambda_{tot} = \lambda_\gamma + \lambda_c$$

where λ_γ is the probability per unit time that the transition will proceed via gamma ray emission and λ_c is the probability per unit time that the transition will proceed via the emission of a conversion electron. The internal conversion coefficient is defined as:

$$\alpha \equiv \frac{\lambda_c}{\lambda_\gamma}$$

where α can have any value between 0 and ∞. (For the $I_i = 0 \rightarrow I_f = 0$ transitions discussed above, $\alpha = \infty$ since there are no gamma rays emitted.) The internal conversion coefficient, α, is the ratio of the total number of conversion electrons emitted per unit time to the number of gamma rays emitted per unit time for a number of identical nuclei making the same nuclear transition.

The decay constant λ_c can be thought of as a composite decay probability which indicates the probability for K, L, M, ... emission — depending upon the electron shell from which the ejected conversion electron originates:

$$\lambda_c = \lambda_K + \lambda_L + \lambda_M + \dots$$

Then:

$$\alpha = \frac{\lambda_c}{\lambda_\gamma} = \frac{\lambda_K + \lambda_L + \lambda_M + \dots}{\lambda_\gamma}$$

$$= \frac{\lambda_K}{\lambda_\gamma} + \frac{\lambda_L}{\lambda_\gamma} + \frac{\lambda_M}{\lambda_\gamma} + \dots$$

$$= \alpha_K + \alpha_L + \alpha_M + \dots$$

where α_K', α_L', α_M', ... are referred to as the K, L, M, ... conversion coefficients.

$$\alpha = \alpha_K + \alpha_L + \alpha_M + \ldots$$

is a rapidly converging series because the outer electrons in an atom have a smaller probability of being inside the nucleus, which decreases the probability of internal conversion.

Experimentally, the various internal conversion coefficients can be measured quite easily. This is possible since the energies of conversion electrons differ as a result of the different binding energies for the various electron shells from which they originate.

Internal conversion coefficients are extremely useful, for they show a strong dependence on the following parameters:

1.   The transition energy, ΔE, between the initial and final nuclear state
2.   The atomic number, Z, of the nucleus
3.   The shell or subshell from which the conversion electron is ejected
4.   The multipolarity, ℓ, or angular momentum of the competing gamma radiation
5.   The character of the nuclear transition, i.e., electric or magnetic

It is important to emphasize that the internal conversion coefficients are essentially independent of nuclear structure. This stems from the fact that both λ_e and λ_γ contain the same matrix elements which thus cancel out in the ratio that is defined as the internal conversion coefficient. Thus $\alpha_K = \lambda_K/\lambda_\gamma$, for example, for a given transition energy, ΔE, and for a given atomic number, Z, depends sensitively on the type and multipolarity of the competing electromagnetic transition. Therefore, when a comparison is made of the experimentally determined internal conversion coefficient and its accurate theoretical value, the angular momentum change and the parity change for a particular nuclear transition can be determined.

In cases where the internal conversion coefficient is large it is often difficult to observe the gamma radiation that is present. In such cases it is advantageous to measure what is called the K/L conversion ratio:

$$\frac{K}{L} = \frac{\alpha_K}{\alpha_L} = \frac{\lambda_K/\lambda_\gamma}{\lambda_L/\lambda_\gamma} = \frac{\lambda_K}{\lambda_\gamma}$$

This ratio can be determined experimentally by observing the relative number of K and L conversion electrons; it is not necessary to observe the associated gamma radiation. In this case also, the nuclear matrix elements cancel out so that the K/L conversion ratio is essentially independent of nuclear structure.

So, by comparing experimental and theoretical values of internal conversion coefficients or K/L conversion ratios, valuable information can be acquired that can be most helpful in making angular momentum and parity assignments for many nuclear states.

REFERENCES

1. Arya, A. P., *Fundamentals of Nuclear Physics,* Allyn and Bacon, Boston, 1966.
2. Beiser, A., *Perspectives of Modern Physics,* McGraw-Hill, New York, 1969.
3. Blatt, J. M., *Theoretical Nuclear Physics,* John Wiley & Sons, New York, 1952.
4. Buttlar, H., *Nuclear Physics,* Academic Press, New York, 1968.
5. Cohen, B. L., *Concepts of Nuclear Physics,* McGraw-Hill, New York, 1971.
6. Eisberg, R., *Quantum Physics of Atoms, Molecules, Solids, Nuclei, and Particles,* John Wiley & Sons, New York, 1974.
7. Enge, H. A., *Introduction to Nuclear Physics,* Addison-Wesley, Reading, Mass., 1966.
8. Evans, R. D., *The Atomic Nucleus,* McGraw-Hill, New York, 1955.
9. Fermi, E., *Nuclear Physics,* rev. ed., The University of Chicago Press, 1950.
10. Fitzgerald, J. J., *Mathematical Theory of Radiation Dosimetry,* Gordon & Breach Science Publishers, New York, 1967.
11. Halliday, D., *Introductory Nuclear Physics,* 2nd ed., John Wiley & Sons, New York, 1955.
12. Kaplan, I., *Nuclear Physics,* 2nd ed., Addison-Wesley, Reading, Mass., 1962.
13. Lederer, C. M., *Table of Isotopes,* 6th ed., John Wiley & Sons, New York, 1967.
14. Leighton, R. B., *Principles of Modern Physics,* McGraw-Hill, New York, 1959.
15. Lipkin, H. J., *Beta Decay for Pedestrians,* North-Holland, Amsterdam, 1962.
16. Livesey, D. L., *Atomic and Nuclear Physics,* Blaisdell, Waltham, Mass., 1966.
17. Meyerhof, W. E., *Elements of Nuclear Physics,* McGraw-Hill, New York, 1967.
18. Miller, D. G., *Radioactivity and Radiation Detection,* Gordon & Breach Science Publishers, New York, 1972.
19. Preston, M. A., *Physics of the Nucleus,* Addison-Wesley, Reading, Mass., 1962.
20. Richtmyer, F. K., *Introduction to Modern Physics,* 6th ed., McGraw-Hill, New York, 1969.
21. Roy, R. R., *Nuclear Physics,* John Wiley & Sons, New York, 1967.
22. Segre, E., *Nuclei and Particles,* W. A. Benjamin, Reading, Mass., 1965.
23. Semat, H., *Introduction to Atomic and Nuclear Physics,* 5th ed., Holt, Reinhart & Winston, New York, 1972.
24. Siegbahn, K., *Beta- and Gamma-Ray Spectroscopy,* North-Holland, Amsterdam, 1955.
25. Weidner, R. T., *Elementary Modern Physics,* 3rd ed., Allyn and Bacon, Boston, 1980.
26. Knolls Atomic Power Laboratory, Chart of the Nuclides, 11th ed. — rev. to April 1972, General Electric Company, Schenectady, New York, 1972.

Chapter 3

INTERACTION OF RADIATION WITH MATTER

Harry T. Easterday

TABLE OF CONTENTS

I.   Introduction 80

II.   X-Ray and γ-Ray Interactions..80
      A.   Introduction 80
      B.   Concept of Cross Section.....................................81
      C.   Photoelectric Effect 83
      D.   Compton Effect 84
      E.   Pair Production 88
      F.   Attenuation and Absorption Coefficients 89

III. Interaction of Heavy Charged Particles with Matter 94

IV.   Interaction of Electrons with Matter 100

V.   Interaction of Neutrons with Matter 102

References...104

I. INTRODUCTION

For the study of interaction with matter, radiations are divided into three groups: (1) electromagnetic radiation (X- and γ-rays), (2) charged particles (electrons, protons, and heavier particles), and (3) uncharged particles (neutrons). The electromagnetic force is dominant in the passage of the first two types of particles through a medium; that is, the incident radiation interacts with the charged constituents of atoms in the absorbing material. At energies greater than 100 MeV, the strong nuclear force begins to compete with the electromagnetic force in the interaction of protons and heavier particles with nuclei. The third type of radiation, neutrons, as the name implies, carry no net charge and lose energy through the nuclear force in collisions with atomic nuclei.

II. X-RAY AND γ-RAY INTERACTIONS

A. Introduction

We consider electromagnetic radiation to have wave-length properties to explain, for example, interference and diffraction effects in optics. Wavelength (λ, cm) and radiation frequency (ν, cycles sec^{-1} or hertz) are related by

$$c = \lambda\nu \tag{1}$$

where $c = 3.00 \times 10^{10}$ cm sec^{-1} (speed of light). In other phenomena (including interaction with matter), electromagnetic radiation behaves like particles (Born,[1] Chapter VI) having energy E and linear momentum p. A connection between the wave and particle aspects is made with the deBroglie relationship

$$E = cp = h\nu = \frac{hc}{\lambda} \tag{2}$$

where $h = 6.63 \times 10^{-27}$ erg sec (Planck's constant). The "particle" of electromagnetic radiation is commonly called a *photon*. The wavelength, λ, can be calculated from the photon energy E by:

$$\lambda(cm) = 0.0124 \times 10^{-8}/E \text{ (MeV)} \tag{3}$$

X-rays and gamma-rays are at the short wavelength (high energy) end of the spectrum of electromagnetic radiation. A distinction between X- and gamma-rays is based on the manner in which the radiation originates. Gamma-rays are by definition restricted to electromagnetic radiation originating from energy changes within the atomic nucleus. X-rays, on the other hand, can be produced in a variety of ways. When atomic electrons make transitions between quantized energy states, X-rays having discrete energies characteristic of each type of atom are emitted and are called *characteristic X-rays*. A continuous X-ray spectrum called *Bremsstrahlung* is produced when electrons free of atoms are deaccelerated by the nuclei of atoms (this effect also occurs but with lesser efficiency with heavier charged particles). Photons emitted in the annihilation of an electron-positron pair are commonly called gamma-rays but are more properly energetic X-rays, often termed *annihilation radiation*.

Photons having energy between a few kiloelectronvolts and 100 MeV interact with matter in any of three competing processes: photoelectric effect, Compton effect, or pair production. Each process will be considered in detail in the following sections and finally their relative importance for various photon energies and absorbing materials will be discussed.

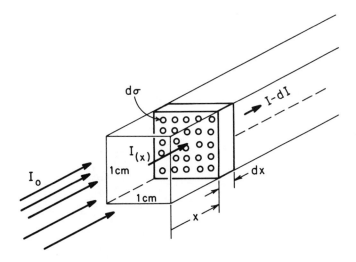

FIGURE 1.   Gamma-ray beam of intensity I_o incident on an absorbing medium.

B. Concept of Cross Section

Without specifying a definite interaction mechanism, consider a number (or intensity I_o) of photons incident on a material. Each atom of the material has a certain probability of interacting with an incident photon; this probability is expressed in the form of a *collision cross section per atom σ* (cm²). Photons are directed along the x-axis and are assumed to form a parallel beam 1 cm² in cross-sectional area (Figure 1). We can now consider each atom as being a disc of area σ (cm²) facing the beam direction. If there are n atoms per cm³ in the medium, the incoming photon beam "sees" n dx atoms (discs) in a thin slab of thickness dx (cm). The probability that a photon will have an interaction in passing through the slab (or the fraction of the incident photons that have an interaction) is equal to the fraction of the 1 cm² area "covered" by the atomic cross sections:

$$\frac{dI}{I(x)} = -(n\,dx)\sigma \tag{4}$$

where $I(x)$ is the photon intensity at a distance x in from the front face of the material and dI is the number of photons removed from the beam in traversing a thickness dx of the material. The cross section σ is usually unrelated to the actual dimensions of the atom, being simply a geometrical way of interpreting the probability of interaction. The frequently encountered *electronic cross section $_e\sigma$* is simply the atomic cross section per electron:

$$_e\sigma = \sigma/Z \tag{5}$$

where Z is the atomic number of the material.

Equation 4 can be solved for $I(x)$ by integration under the assumption that I_o photons are incident on the front face (x = 0):

$$I(x) = I_0 e^{-n\sigma x} \tag{6}$$

The reader can convince himself the formula is valid for a beam of any cross-sectional

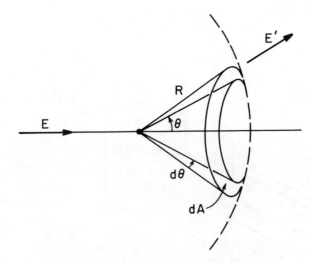

FIGURE 2.   Photon of energy E scattered between angles θ
and $\theta + d\theta$.

area, the only restriction being that the area of the material must be large enough to
intercept all of the incident photon beam. When several different types of processes
can remove photons from the beam, a cross section for each process may be defined.
The total cross section for any interaction is then the sum of the individual cross sections. The intensity relationship of Equation 6 is valid only for the total collision cross
section (see Section II.F).

Often it is of interest to know the probability that the photon will be scattered at
some specified angle after the collision. In many cases of interest the only angular
dependence of the scattering is on the angle θ with respect to the x-axis (Figure 2).
Suppose a sphere of radius R is drawn about the atoms and we observe the number of
photons emerging between angles θ and $\theta + d\theta$. This region of space is termed the
solid angle $d\Omega$ and is defined as the area dA on the sphere enclosed between angles θ
and $\theta + d\theta$ divided by the square of the radius:

$$d\Omega = \frac{dA}{R^2} = \frac{2\pi R \sin\theta \; R \; d\theta}{R^2} = 2\pi \sin\theta \; d\theta \qquad (7)$$

The *differential scattering cross section per atom* $d\sigma(\theta)$ is the probability that an incident photon will be scattered into solid angle $d\Omega$ at scattering angle θ per atom per
square centimeter of material normal to the beam.

A fundamental assumption in these developments is that when a photon does interact in some manner, either by being absorbed or scattered by the atom, the photon is
removed from the beam. Such is not always the case with radiation other than photons.
For example, the largest fractional energy loss that can occur in the collision (head-
on) of a heavy charged particle of mass M with an electron of mass m_o is $4m_o/M$. Since
heavy charged particles are several thousand times more massive than electrons, thou-
sands of collisions must be made by a heavy charged particle in its passage through a
substance before the incident particle loses all of its original kinetic energy. Except for
statistical fluctuations, all particles in a monoenergetic beam of heavy particles travel
the same distance, called the *range*, before coming to rest. Thus the number of particles
comprising the beam remains constant and the initial direction is maintained during
the slowing down process. In contrast, a photon is removed from the beam in one
interaction. This "one shot" characteristic of the photon interaction results in an ex-

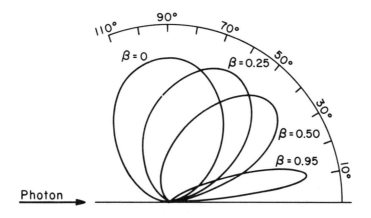

FIGURE 3.   Angular distribution of photoelectrons, β = v/c where v is the velocity of the photoelectron. Distributions are symmetric around photon direction.

ponentially decreasing beam intensity (Equation 6). The concept of a photon "range" has no meaning since photons are removed from the beam at widely varying distances into the material. Instead, the "half-value thickness" (the amount of absorber required to decrease the beam intensity by one half) is a measure of the ability of a material to absorb photons.

C. Photoelectric Effect

In the *photoelectric effect,* the photon is absorbed by a bound electron giving up all its energy and in effect vanishing. The electron ("photoelectron") in turn is ejected from the atom, carrying away the energy of the incident photon less the energy required to free the previously bound electron from the atom:

$$T = E - B_e \qquad (8)$$

where T is the kinetic energy of the photoelectron, E is the energy of the photon, and B_e is the binding energy. Photoelectrons ejected from a given atomic shell by photons of a fixed energy are monoenergetic even if emitted at differing directions by the absorbing atoms. The residual atom, left with a vacancy in one of its electron shells emits characteristic X-rays or Auger electrons in returning to its equilibrium state.

It can be shown using the laws of conservation of momentum and energy that a photon cannot give up its entire energy to a free electron. However, when the electron is bound to the nucleus of an atom, the entire atom is able to participate in the interaction, thus permitting complete absorption of the photon to occur. Since the atom is so massive compared to an electron, recoil energy acquired by the atom in the momentum balance can be neglected. The greater the binding energy of the electrons, the greater is the probability of photon absorption. Hence, the photoelectric effect is enhanced whenever (1) an inner shell (usually K shell) electron is involved; (2) the atomic number Z of the nucleus is high; and (3) the energy of the photon is relatively low.

The photon energy must exceed the binding energy of the photoelectron if the electron is to be ejected with positive kinetic energy. As the photon energy is increased above one of the electron shell binding energies, an additional set of more tightly bound electrons becomes available for the photoelectric process. Thus sharp increases in the photoelectric cross section (called K- and L-edges) occur at the corresponding shell binding energies, B_e. A tabulation of binding energies may be found in Volume I, Chapter 2.

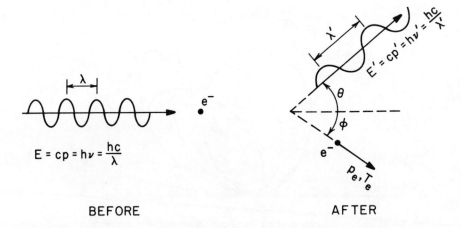

$$E = cp = h\nu = \frac{hc}{\lambda}$$

BEFORE   AFTER

FIGURE 4.   Compton effect: collision of a photon of energy E with a free electron. E′ is the energy of the scattered photon and β = v/c.

No simple expression is adequate for the calculation of photoelectric cross sections. For photon energies beyond the absorption edges the cross section per atom is approximately proportional to Z^4E^{-3}. Weber and Van den Berge[21] found that the empirical formula

$$\sigma_{Ph} = 17.7 \times 10^{-24}\ Z^{4.4}\ E^{-3.1}\ cm^2/atom \qquad (9)$$

(E in keV) is accurate to within 5% for elements of atomic number less than 20 and photon energies below 200 keV.

A plot of the quantity μ_{Ph} which is proportional to μ_{Ph} is shown in Figures 7 to 9 for several elements.

Photoelectrons tend to be ejected perpendicular to the beam direction by low energy photons and in a more forward direction by more energetic photons (see Figure 3). The differential cross section $d\sigma$ for the ejection of a photoelectron into solid angle $d\Omega$ at angle θ relative to the direction of the photon is proportional to

$$d\sigma(\theta) \propto \frac{\sin^2\theta\ d\Omega}{\left(1 - \dfrac{v}{c}\cos\theta\right)^4} \qquad (10)$$

for photoelectrons having energies up to several hundred keV.   Here, v/c is the fraction of the electron velocity relative to that of light.

D. Compton Effect

The process in which a photon collides with an electron and a scattered photon of lower energy is produced is called the *Compton effect.*

Compton scattering is usually important only when the incident energy of the photon is large compared to the binding energy of the electron in the atom. When the photon energy is comparable to the binding energy, the photoelectric effect predominates. Hence, in Compton scattering we can consider all electrons in the medium to be free and at rest. The incident photon is taken to be a "particle" of energy E and linear momentum p (see Figure 4). By using the concepts of conservation of energy and vector linear momentum, formulas for the energy of the scattered photon and electron can be derived.

A photon of incident energy E scattered through angle θ has scattered energy E' given by

$$E' \doteq \frac{E}{\alpha(1 - \cos\theta) + 1} \tag{11}$$

where $\alpha = E/m_o c^2$ and $m_o c^2 = 0.511$ MeV is the rest energy of an electron.

The kinetic energy T of the scattered electron is

$$T = E - E' \tag{12}$$

$$T = \frac{E\alpha(1 - \cos\theta)}{1 + \alpha(1 - \cos\theta)} \tag{13}$$

If the electron is scattered through an angle ϕ (Figure 4), the relationship between angles θ and ϕ is

$$\cot\phi = (1 + \alpha) \tan \frac{\theta}{2} \tag{14}$$

We note from Equation 11 that (1) the energy of the scattered photon depends only on the angle of scatter and the incident energy of the photon; (2) the greater the angle of scatter, the smaller is the energy of the scattered photon; and (3) the photon cannot lose all of its energy in a scattering event. The minimum scattered photon energy (and maximum electron recoil energy) occurs when $\theta = 180°$. Using this value in Formulas 11 and 13 yields

$$E'_{min} = m_o c^2 \left(\frac{\alpha}{1 + 2\alpha} \right) \tag{15}$$

$$T_{max} = E \left(\frac{2\alpha}{1 + 2\alpha} \right) \tag{16}$$

It is interesting to note that as the incident photon energy becomes large ($E \gg m_o c^2$), the backscatter ($\theta = 180°$) photon energy approaches a constant value $E' = m_o c^2/2 = 0.25$ MeV. The accompanying recoil electron must move in the forward direction ($\phi = 0°$) and its energy is approximately that of the incident photon. At low energies ($E \ll m_o c^2$) energy loss in scattering becomes negligible for all scattering angles, and the electron recoil energy is essentially zero.

Unless the photon beam is polarized, the number of scattered photons and electrons, although dependent on the angle of scattering θ, will not depend on the azimuthal angle measured around the direction of the incident beam. Then the differential scattering cross section per electron for photons incident on free electrons is given by the Klein-Nishina formula:[3]

$$\frac{d_e \sigma(\theta)}{d\Omega} = \frac{r_o^2}{2} \left(\frac{E'}{E} \right)^2 \left(\frac{E}{E'} + \frac{E'}{E} - \sin^2\theta \right) \tag{17}$$

where

$$r_o = \frac{e^2}{m_o c^2} = 2.82 \times 10^{-13} \text{ cm} \tag{18}$$

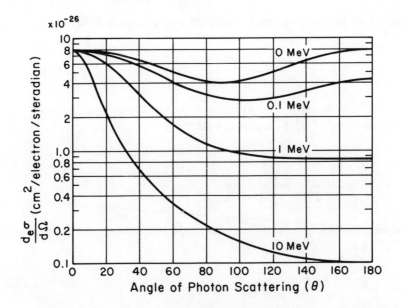

FIGURE 5.   Variation of Compton differential scattering cross section (per electron) $d_e\sigma(\theta)/d\Omega$ with photon scattering angle θ.

is the classical electron radius, and e is the electric charge. A plot of this formula for various photon energies is shown in Figure 5.

In the limit of low photon energies, $E' \cong E$ and Equation 17 becomes

$$\frac{d_e\sigma(\theta)}{d\Omega} = \frac{r_o^2}{2}\ (1 + \cos^2\theta) \qquad (19)$$

The expression implies that the distribution of scattered photons is symmetric about $\theta = 90°$ (see Figure 5). This result can be understood from the following: at low frequencies the free electron is forced by the electric field of the incident photon to oscillate along a line perpendicular to the path of the photon. The oscillating electron re-radiates electromagnetic energy of the same frequency as the incident photon ($E' = E$) and with a $\cos^2\theta$ "donut without a hole" distribution emitted by dipole radio antennas. The pattern is symmetric about 90° with no radiation emitted at $\theta = 90°$. The total radiation is a result of many incident unpolarized photons which will yield the $(1 + \cos^2\theta)$ dependence given above.[4]

The total cross section for the case of low photon energy can be found by integration:

$$_e\sigma = \int_{4\pi} \left(\frac{d_e\sigma(\theta)}{d\Omega}\right)\ d\Omega = \frac{8\pi}{3}\ r_o^2 = 0.665 \times 10^{-24}\ \text{cm}^2/\text{electron}$$

$$(20)$$

The most important features of low energy scattering are the absence of an energy shift characteristic of Compton scattering and the "fore-aft" symmetry of the radiation pattern. This process is given the name *Thomson scattering*.

Total cross section for any photon energy is obtained from integration of Equation 17:

$$_e\sigma = 2\pi r_o^2 \left\{ \frac{1+\alpha}{\alpha^2} \left[\frac{2(1+\alpha)}{1+2\alpha} - \frac{1}{\alpha}\ \ell n(1+2\alpha) \right] + \frac{1}{2\alpha}\ \ell n(1+2\alpha) - \frac{1+3\alpha}{(1+2\alpha)^2} \right\}\ \text{cm}^2/\text{electron} \qquad (21)$$

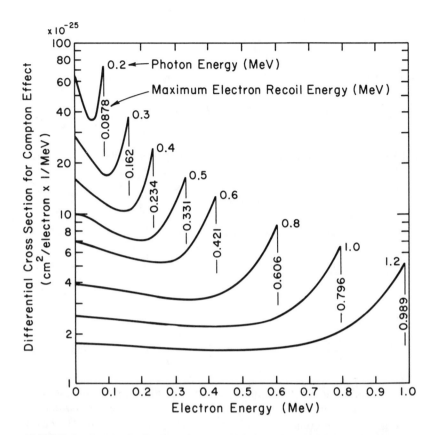

FIGURE 6.   Energy distribution of scattered electrons from Compton effect (in the form of a differential cross section) for various incident photon energies.

This relation gives the probability per electron per square centimeter of material that an incident photon will be scattered. A plot of the proportional quantity μ_{c_o} is given in Figures 7 to 9.

Other relations that are often useful are the following (see Attix and Roesch,[5] chap. 2, Evans,[6] chap. 23):

1.   Probability per unit energy that an incident photon will produce a Compton electron having kinetic energy between T and T + dT:

$$\frac{d_e\sigma}{dT} = \frac{2\pi\, m_0 c^2}{(E-T)^2} \frac{d_e\sigma}{d\Omega} \qquad \frac{cm^2}{keV \cdot electron} \qquad (22)$$

The energy distribution of scattered electrons for various incident photon energies is given in Figure 6.

2.   Probability per unit energy that an incident photon will produce a scattered photon having energy between E′ and E′ + dE′:

$$\frac{d_e\sigma}{dE'} = \frac{2\pi\, m_0 c^2}{E'^2} \frac{d_e\sigma}{d\Omega} \qquad \frac{cm^2}{keV \cdot electron} \qquad (23)$$

3.   Probability per unit solid angle that a unit of incident photon energy is scattered into solid angle dΩ at scattering angle θ:

$$\frac{d\ _e\sigma_s}{d\Omega} = \frac{E'}{E} \frac{d\ _e\sigma}{d\Omega} \quad \frac{cm^2}{\text{steradian electron}} \tag{24}$$

4. Probability that a unit of incident photon energy is scattered into any angle:

$$_e\sigma_s = \int_{4\pi} d\ _e\sigma_s = \pi r_o^2 \left[\frac{1}{\alpha^3} \ln(1+2\alpha) + \frac{2(1+\alpha)(2\alpha^2 - 2\alpha - 1)}{\alpha^2 (1+2\alpha)^2} + \frac{8\alpha^2}{3(1+2\alpha)^3} \right] \frac{cm^2}{\text{electron}} \tag{25}$$

5. Probability that a unit of incident photon energy is absorbed by an electron:

$$_e\sigma_a = \ _e\sigma - \ _e\sigma_s \quad \frac{cm^2}{\text{electron}} \tag{26}$$

6. Average energy of a scattered photon:

$$E'_{av} = \frac{_e\sigma_s}{_e\sigma} E \tag{27}$$

7. Average kinetic energy of a Compton electron:

$$T_{av} = \frac{_e\sigma_a}{_e\sigma} E \tag{28}$$

The assumption made at the beginning of this section that we consider electrons to be "free" is invalid when photon energies are low enough to be comparable to electron binding energies. This is roughly 10 keV in the region of oxygen and 100 keV for high-Z elements such as lead. In this situation the photon is considered to be a wave that interacts with all of the electrons as it passes through an atom. Electrons affected by the wave then act as coherent sources of secondary electromagnetic waves which add together to form the outgoing scattered wave. The intensity of this *coherently scattered radiation* is greater than that predicted by the preceding equations since the photon had been considered to interact with only a single electron. Since this effect occurs at relatively low photon energies where the photoelectric effect is the dominant interaction process, errors introduced by neglecting coherent Compton scattering amount to a few percent at most.

E. Pair Production

A photon can be absorbed in the field of an atomic nucleus resulting in the creation of an electron and positron. This process is called *pair production*. The positron ultimately annihilates with an electron resulting in the emission of two 511 keV photons. When the pair is created, no net electric charge is generated since the charge carried by the positron is the same in magnitude but opposite in sign to that of the electron. However, mass has been created; pair production is an example of the direct conversion of energy into mass (E to mc^2) so the photon must have energy equivalent to the rest energy of the pair, $2m_oc^2 = 1.02$ MeV, for the process to occur. Photon energy in excess of this minimum is shared (not necessarily equally) between the electron and positron as kinetic energy. The only (but necessary) function of the nucleus is to provide the means of conserving linear momentum.

Pair production can also occur in the vicinity of an electron and is called triplet production. Being much lighter than an atom, an electron will recoil with a comparatively greater kinetic energy. The minimum photon energy for pair production in the field of an electron is $4m_oc^2 = 2.04$ MeV. In the competition between several possible

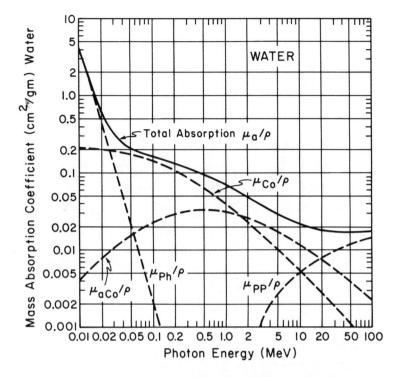

FIGURE 7. Gamma-ray mass absorption coefficient for water.

reactions, in general the most probable is that reaction in which the largest fraction of the incident energy is available to the products. Pair production in the field of electrons is less probable than in the field of nuclei because the recoil electron must acquire a greater amount of the available energy than a recoil atom.

The ratio of triplet to pair production per atom for a material of atomic number Z is approximately given by the relation:[6]

$$\frac{\sigma_{\text{trip}}}{\sigma_{\text{pair}}} = \frac{1}{C\,Z} \tag{29}$$

where C varies from a value of about 3 for a photon energy of 6 MeV to unity as the photon energy becomes infinitely large.

The pair production cross section σ_{pp} is proportional to Z^2:

$$\sigma_{pp} = k(E)\,Z^2 \tag{30}$$

k(E) being a complicated function of the photon energy E. A plot of the quantity μ_{pp} which is proportional to σ_{pp} is shown in Figures 7 to 9 as a function of photon energy.

F. Attenuation and Absorption Coefficients

Although the ability of an atom to absorb or scatter photons can be expressed in the form of a cross section, it is more usual to employ the concept of *attenuation coefficient*. The *linear attenuation coefficient* μ is defined as

$$\mu = n\sigma = n\,Z_e\sigma \quad (\text{cm}^{-1}) \tag{31}$$

where n, the number of atoms per volume of material, can be found from Avogadro's

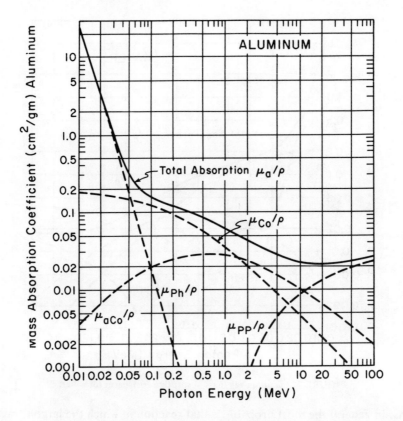

FIGURE 8.   Gamma-ray mass absorption coefficient for aluminum.

number N_A, the atomic mass A (gram per molecule) of the material, and the density ϱ as

$$n = \frac{\rho\, N_A}{A} \qquad (32)$$

Equation 4 then becomes

$$\frac{dI}{I(x)} = -\mu\, dx \qquad (33)$$

Thus μ is the probability of a photon interacting with the material in traveling a small distance dx. Integrating, we have for the photon intensity at a distance x into the material

$$I(x) = I_o\, e^{-\mu x} \qquad (34)$$

where I_o is the intensity at x = 0. The reciprocal of μ is often termed the *mean path length* since it can be shown to be equal to the mean distance a photon travels before interacting in the material. The thickness of material required to remove one half of the photons from the beam is called the *half value layer* HVL and is given by

$$HVL = 0.693/\mu \qquad (35)$$

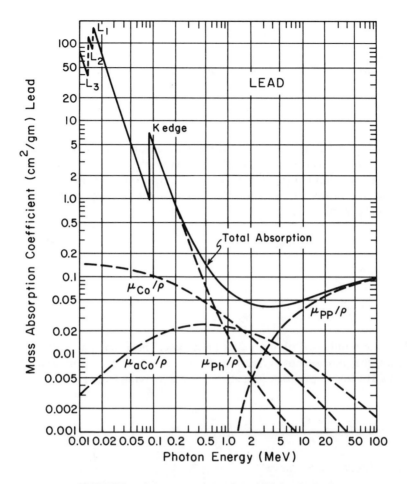

FIGURE 9. Gamma-ray absorption coefficient for lead.

It is also common in the literature to express the thickness (or better the "mass thickness") as t (g cm^{-2}), and since x (cm) and t are related as follows:

$$t = \rho x \tag{36}$$

(where ϱ is the density of the medium in g cm^{-3}) then a mass attenuation coefficient can be defined from

$$\frac{\mu}{\rho} t = \mu x \tag{37}$$

Justification for introducing mass absorption μ/ϱ with its rather obscure unit, g cm^{-2}, in place of the more obvious cm^{-1}-unit of linear absorption is as follows: attenuation of a photon beam depends upon the number of electrons encountered in traveling a unit distance in the material, which is given by

$$Zn = \frac{Z N_A \rho}{A} \tag{38}$$

Although densities may vary over several orders of magnitude for various materials, Z/A changes by only 20% throughout the periodic table (neglecting hydrogen). Thus photon attenuation is approximately proportional to the density of the material. Mass attenuation coefficients for differing substances tend to be more alike than the corresponding linear coefficients, and, being independent of the density of the material, are in addition more expedient to tabulate for any given substance. Equation 34 becomes

$$I(x) = I_o \, e^{-\frac{\mu}{\rho} t} \tag{39}$$

Cross sections (and attenuation coefficients) measure the probability of occurrence of an absorption or scattering event. Since a given photon can be involved in at most one of the three competing processes (photoelectric effect, Compton effect, or pair production), the probability of *any* event occurring is the sum of the probabilities of the individual events:

$$\sigma_T = \sigma_{Ph} + \sigma_{Co} + \sigma_{pp} \quad (\text{cm}^2) \tag{40}$$

so

$$\frac{\mu_T}{\rho} = \frac{\mu_{Ph}}{\rho} + \frac{\mu_{Co}}{\rho} + \frac{\mu_{pp}}{\rho} \quad (\text{cm}^2 \text{ g}^{-1}) \tag{41}$$

To find the number of photons that have experienced one particular kind of interaction (say, the photoelectric effect) in going a thickness t, we note that the intensity of photons dI_{Ph} lost to photoelectric absorption in moving from t to t + dt is

$$dI_{Ph} = \frac{\mu_{Ph}}{\rho} I \, dt = \frac{\mu_{Ph}}{\rho} I_o \, e^{-\frac{\mu_{Ph}}{\rho} t} \, dt \tag{42}$$

Integrating from t = 0, we have

$$\frac{I_{Ph}}{I_o} = \frac{\mu_{Ph}}{\mu_T} \left(1 - e^{-\frac{\mu_T}{\rho} t} \right) \tag{43}$$

Similar expressions hold for Compton scattering and pair production. It is assumed that t is small enough that multiple scattering does not occur.

Mass attenuation coefficients for photons incident on water, aluminum, and lead are shown in Figures 7 to 9. Certain features of these curves should be noted. The photoelectric effect is dominant in the low energy region (where the absorption edges occur), below 50 keV for aluminum and 500 keV for lead. Absorption coefficients for the photoelectric effect decrease rapidly with increasing photon energy above the absorption edges. The Compton effect then becomes the principal interaction mechanism, dominating over a broad energy region: 60 keV to 10 MeV in aluminum and 70 keV to 4 MeV in lead. Absorption of photons above these limits is primarily by pair production.

Numerical values of mass attenuation coefficients, μ/ϱ, are tabulated in the literature (e.g., Attix and Roesch[5]) for a series of photon energies for each element. The following procedure can be used to calculate attenuation coefficients when the absorbing material is made up of several elements, either as a chemical compound or as a simple mixture of separate elements. The mass attenuation coefficient for the mixture is

$$\frac{\mu}{\rho} = \frac{\mu_1}{\rho_1} f_1 + \frac{\mu_2}{\rho_2} f_2 + \ldots \ldots \tag{44}$$

where f_1, f_2, etc. are the fractions by *weight* for the elements designated 1, 2, ...(e.g., for H_2O, $f_H = 1/9$ and $f_O = 8/9$).

Equations 34, 40, and 43 are used to calculate the *number* of photons (or number per second) which interact with the absorbing medium. In calculating the dose (or dose rate) delivered to a biological material by a beam of photons, the *energy* deposited is of importance. A photon of energy E_o(MeV) deposits $E_o - 2m_oc^2$ (see Section II.E) for each pair production interaction and $E_o - B_e$ (Equation 8) for each photoelectric interaction; in each interaction the incident photon disappears and the energy of the resulting electrons is absorbed by the material. Since the Compton effect consists of a scattering event, part of the incident photon energy is delivered to the recoil electron (which is absorbed) and the remaining energy leaves the interaction region as a photon of lower energy. Thus each interaction deposits a lesser amount of energy than the full photon energy; the *attenuation* coefficient μ_{Co} (Equation 41) must be replaced with a smaller *absorption* coefficient μ_{aCo} in calculations for energy absorption.[5] Under the assumption that the terms $2m_oc^2$ and B_e can be neglected, the total *absorption* coefficient then becomes

$$\mu_a = \mu_{aCo} + \mu_{Ph} + \mu_{PP} \quad (cm^{-1}) \tag{45}$$

If I is the number of photons per square centimeter per second in a beam, then the number/second removed from the beam in traversing a thickness dx (cm) of the absorber is

$$dI = -\mu I \, dx \quad (photons/cm^2 \, sec)$$

If each photon has energy E_o (MeV), then the incident energy intensity is

$$I_E = E_o I \quad (MeV/cm^2 \, sec)$$

and the energy absorption in a thickness dx (cm) is

$$dI_E = -\mu_a I_E dx \quad (MeV/cm^2 \, sec)$$

The energy absorption per unit volume becomes

$$\frac{dI_E}{dx} = -\mu_a I_E \quad (MeV/cm^3 \, sec) \tag{46}$$

The dose rate per gram can be found by dividing by ϱ, the density of the material

$$\frac{1}{\rho} \frac{dI_E}{dx} = -\left(\frac{\mu_a}{\rho}\right) I_E \quad (MeV/gsec) \tag{47}$$

A very complete discussion of mass attenuation and absorption coefficients (including tables) is in Hubbell.[7]

More details of the physics of the interaction of radiation with matter may be found in Evans,[6] and Marmier et al.[10] Heitler[3] is the standard reference for the theory of the subject. Readers who are interested in dosimetry are referred to Hine and Brownell[2] and Attix and Roesch.[5] Useful tables of cross sections, attenuation and absorption coefficients can be found in Hubbell,[7,8] Hine and Brownell,[2] Attix and Roesch,[5] Gray,[9] and Grodstein.[11] An extensive bibliography of atomic and nuclear data used in biomedical applications is given in Holt.[12]

III. INTERACTION OF HEAVY CHARGED PARTICLES WITH MATTER

The paths of heavy charged particles (protons, deuterons, alpha particles, etc.) can be observed with the aid of cloud, bubble chambers, and photographic emulsions. With few exceptions the tracks are staight and show little deflection during the slowing-down process. This would indicate that energy loss is by interaction with electrons (excitation and ionization of the absorbing material) rather than by collisions with more massive nuclei since large deflections would occur in the latter case. The rarity of nuclear collisions in comparison to interaction with atomic electrons is simply a size effect; nuclei are roughly 10^{-13} cm in radius whereas the electron clouds which make up the atom are much larger, around 10^{-8} cm in radius. Electronic excitation consists of an outer-shell electron being forced into a higher unoccupied energy orbit. In the ionization process, an atomic electron (not necessarily an outer-shell electron) is separated from the atom which then becomes a positively charged ion.

Photographs of the paths of heavy charged particles show, in addition to the prominent straight trace, many much lighter tracks branching off from the main path and extending short distances into the medium. It is possible for an electron to gain sufficient energy from the heavy particle in the original ionization process to in turn cause secondary ionizations before coming to rest. The fainter tracks are produced by these energetic electrons (called "delta rays").

Energy loss in the collision of a heavy charged particle (charge $+ze$ and rest mass M_o) with an electron (charge $-e$ and rest mass m_o) can be understood as follows: the largest energy transfer occurs from a "head-on" collision where the fractional energy change for the heavy particle is $4m_o/M_o$. The electron to proton mass ratio is $m_o/M_o = 1/1836$, so the maximum possible energy loss per collision is small even for the lightest of the "heavy" particles. Energy loss for a typical encounter is small enough so the initial speed, v, of the heavy particle can be considered unchanged in a collision. The electron, regarded to be initially free and at rest, acquires sidewise momentum in the collision as a result of the attractive electrostatic force exerted by the heavy particle as it moves past (see Figure 10). Let b be the closest separation distance in the collision (called the impact parameter) and x measures the instantaneous location of the heavy particle, ze, from the point 0. The sidewise momentum acquired by the electron is

$$\Delta p_\perp = \int_{-\infty}^{+\infty} F_\perp \, dt = \int_{-\infty}^{+\infty} F(\cos\theta) dt = \int_{-\infty}^{+\infty} \left(\frac{ze^2}{x^2 + b^2} \right) \left(\frac{b}{\sqrt{x^2 + b^2}} \right) \frac{dx}{v} = \frac{2ze^2}{bv} \tag{48}$$

Energy gained by the electron (from the loss by the heavy particle) is then

$$\Delta E_b = \frac{(\Delta p_\perp)^2}{2m_o} = \frac{2z^2 e^4}{m_o v^2 b^2} \tag{49}$$

ΔE_b represents the amount of energy lost by the heavy particle in a single collision of impact parameter b. In order to calculate the energy loss in going a distance dx in the medium, we sum the energy losses for all electrons encountered. Each electron within a cylindrical shell of radius b, shell thickness db, and length dx gains energy ΔE_b, so the energy gain for all electrons in the cylindrical shell is

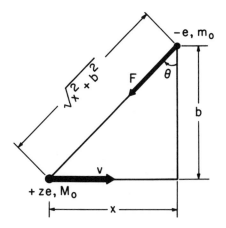

FIGURE 10. Electrostatic force F exerted by a heavy particle ($+ze$) as it moves past an electron ($-e$); impact parameter is b.

$$dE_{b+db} = 2\pi b \, (db) \, (dx) \, NZ\Delta E_b \qquad (50)$$

where N is the number of atoms per unit volume of the medium and Z is the number of electrons per atom. This expression is then integrated over all possible impact parameters to get the total energy loss, dE, in going a distance, dx:

$$-\frac{dE}{dx} = \frac{4\pi z^2 e^4 NZ}{m_0 v^2} \int_{b_{min}}^{b_{max}} \frac{db}{b} = \frac{4\pi z^2 e^4 NZ}{m_0 v^2} \ln \frac{b_{max}}{b_{min}} \qquad (51)$$

The lower and upper limits of the impact parameter, b_{min} and b_{max}, must be finite because of the $\ln b$ dependence. In the frame of reference of the heavy particle, the approaching electron appears to be a wave. Since the location of the electron can only be specified to within a deBroglie wavelength (Equation 2), $b_{min} \propto 1/m_0 v$. The quantity b/v is a measure of the time duration of the impact; that is, the larger b and the smaller v, the longer the collision lasts. No net energy is transferred to the electron in a collision of duration comparable to the time it takes a bound electron to make a complete orbit around the atom. The latter time is inversely proportional to ionization potential (or binding energy), I, since the more tightly bound an electron is the faster it moves. (Or more crudely put: in an impact of long duration, the heavy particle causes the electron to speed up in the approaching phase of its orbit and to slow down in the receding phase in equal amounts.) The energy gained by the electron is small at large impact parameters ($\Delta E_b \propto b^2$ from Equation 49). It is reasonable that the upper bound of the impact parameter, b_{max}, should depend on the ionization potential since an electron can only absorb energy if the amount is sufficient either to raise the electron to an unoccupied state (excitation) or to free it from the atom (ionization).

A more detailed derivation involving quantum mechanical and relativistic effects[13] yields

$$-\frac{dE}{dx} = \frac{4\pi z^2 e^4 NZ}{m_0 v^2} \left[\ln\left(\frac{2m_0 v^2}{\overline{I}(1-\beta^2)}\right) - \beta^2 - \sum_i C_i/Z \right] \quad \text{(ergs/cm)} \qquad (52)$$

where $\beta = v/c$ and $\overline{I}$ is a suitably averaged ionization potential. This relationship is

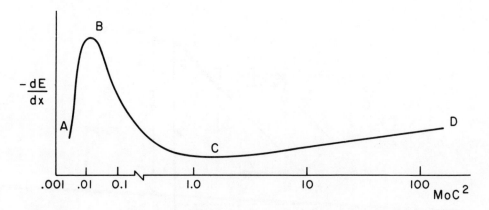

FIGURE 11. Energy loss rate dE/dx as a function of kinetic energy for a particle of rest mass-energy $M_o c^2$.

commonly called the *Bohr stopping power* formula and is the basis for the theory of charged particle energy loss. When the energy of the heavy particle has been reduced to a value comparable to the electron binding energy of a particular atomic shell, these electrons can no longer participate in the energy loss process. The term $\Sigma \, C_i/Z$ in Equation 52 is the "shell correction factor" which accounts for this effect.[5,6] The index i designates the shell, that is, i = K, L, etc.

General features of this formula are illustrated in Figure 11. The horizontal scale is the kinetic energy of the heavy particle expressed as a product of its rest energy, $M_o c^2$, (938 MeV for a proton). The gradual increase at extremely high energies, C to D, results from the $(1 - \beta^2)$ factor in the denominator. At low velocities (v $\ll$ c) the electric field intensity of a charged particle has spherical symmetry about the particle. At relativistic speeds (v $\rightarrow$ c), the electric field becomes stronger in the directions transverse to the particle path. This results in a slight increase in b_{max} over its value at lower speeds and a larger dE/dx. The point C is the energy for minimum energy loss rate (roughly $M_o c^2$). At low speeds positively charged ions tend to attach electrons resulting in a lower net charge and decreased dE/dx shown in the region AB.[5] A helium ion, for example, doubly charged at high energies (z = 2) is singly charged for roughly half the time at 0.8 MeV.

Given the kinetic energy T(MeV) of a particle of charge ze, mass m_o (grams) and rest energy $m_o c^2$ (MeV) (note: 1 AMU = 1.66×10^{-24} g = 931 MeV, e = 4.8×10^{-10} esu), the following procedure can be used to calculate each term of the formula for stopping power, dE/dx (MeV/cm) (Equation 52).

1.   N = number of atoms per centimeter of the absorbing material:

$$N = \frac{N_A \rho}{A}$$

where N_A = 6.02×10^{23} atoms/mole (Avogadro's number), A = atomic number (grams/mole) of the absorbing material, ϱ = density of the absorber (g/cm³).

2.

$$\beta = \frac{v}{c} = \sqrt{1 - \left(1 + \frac{T}{m_o c^2}\right)^{-2}}$$

where v = particle speed (cm/sec), c = 3.00×10^{10} cm/sec (speed of light).

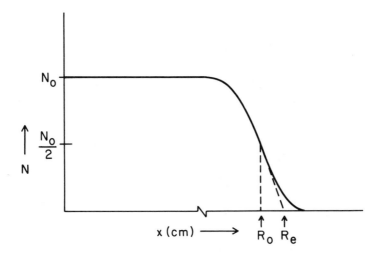

FIGURE 12. Number- distance curve for 6 MeV alpha- particles in air; the effect of straggling has been exaggerated, $R_e - R = 0.1$ cm. $R = 4.7$ cm.

3.   A good approximation for the mean ionization potential is given by Evans[6]

$$\bar{I} \simeq 1.84 \times 10^{-11} \, Z \, \text{(ergs)}$$

where Z is the atomic number of the absorber.

4.   Graphs of the shell correction factor C_i/Z are given in Evans,[6] and Attix and Roesch.[5]

5.   To convert from stopping power (ergs/cm) to (MeV/cm) divide by 1.60×10^{-6}.

As can be seen from Figure 11, the stopping power decreases slowly with increasing particle energy between B and C (a few keV to 1000 MeV for protons). The stopping power in this region has the functional form:

$$dE/dx = k \, T^{-\alpha} \tag{53}$$

where the terms k and α vary slowly with kinetic energy, T. If stopping powers are known for two values of T, k and α can be calculated and the formula can then be used to find dE/dx for other values of T. The procedure is in essence an interpolation method.

The chemical state of an element does not appreciably affect its stopping power. In computing the stopping power of a chemical compound or mixture, dE/dx must be calculated for each element and the mean stopping power is the sum:

$$\overline{(dE/dx)} = f_1 (dE/dx)_1 + f_2 (dE/dx)_2 + \ldots \tag{54}$$

where f_1, f_2, etc., are atomic *number* fractions for elements designated 1, 2, etc. (e.g., for H_2O, $f_H = 2/3$ and $f_o = 1/3$).

A number-distance curve for heavy charged particles is shown in Figure 12. As the particles, N_o in number and all of the same initial energy, penetrate into the absorber, the number in the beam does not change since energy loss is a gradual process and collisions with electrons do not appreciably alter the path direction. Energy loss is not identical for each particle, however, and the total distance a particle must go to lose all of its energy varies by a slight amount about the "mean range", R. The *extrapolated range*, R_e, can be found from the number-distance curve by drawing a straight

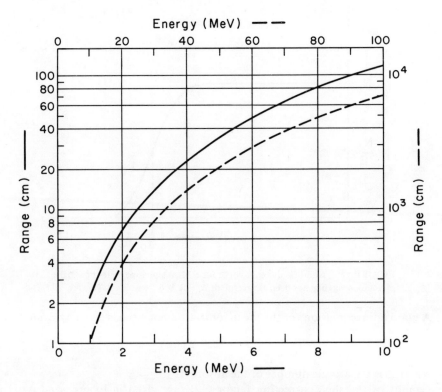

FIGURE 13.   Range-energy for protons in air.

line tangent to the curve at its steepest slope to the distance axis. The difference between the two ranges, $R_e - R$, called *straggling* amounts to less than 2% of the range.[14] It should be possible to derive the mean range for a particle of energy E_o by integrating the stopping power formula:

$$R = \int_{E_o}^{0} \frac{dE}{-(dE/dx)} \qquad (55)$$

Uncertainty in how to correct for the change in the charge of the heavy particle at low energies (region AB of Figure 11) makes this procedure inaccurate. Note that from the form of the stopping power formula the rate of energy loss depends on the speed and charge of the heavy particle but not on its mass (e.g., a proton and a deuteron of the same velocity have the same energy loss rate; a doubly charged helium ion of the same velocity would have four times the energy loss rate). Thus, knowing the range-energy relationship for protons ($z = 1$, $M_p = 1$ AMU) in a certain medium, the range of any particle of charge ze and mass M, having the same initial velocity as the proton, can be found using

$$R(Z,M) = \frac{1}{z^2} \frac{M}{M_p} R_p \qquad (56)$$

Particles having the same initial velocity have initial kinetic energies to T_o ($T_o \ll m_o c^2$) in the same ratio as their masses:

$$\frac{T_o(M)}{T_o(p)} = \frac{M}{M_p}$$

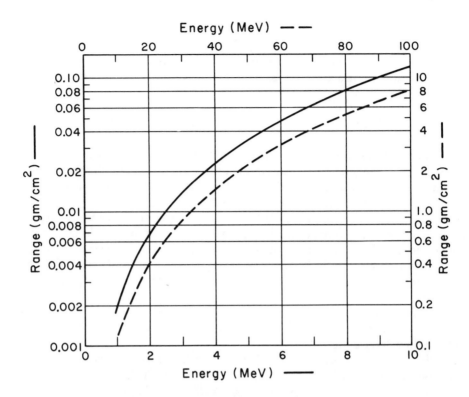

FIGURE 14.   Range-energy for protons in water.

Hence,

$$R(Z,M) = \frac{1}{z^2} \frac{T_0(M)}{T_0(p)} R_p \qquad (57)$$

For example, the range of a 12 MeV α-particle (helium ion, $Z = 2$, $M = 4.0$ AMU) in air is 14.2 cm; this is also the range of a 3 MeV proton.

Proton range - energy curves for air and water are given in Figures 13 and 14.

If the range in a medium of a certain particle of given kinetic energy is known, how can the range of this particle in a different medium be calculated? It is not obvious from the preceding formulas but is known to be true from experiment that

$$R \propto \frac{\sqrt{A}}{\rho}$$

where A and ϱ are the atomic weight and density, respectively, of the material.

This proportionality is known as the *Bragg-Kleeman rule:* it can be written as a ratio to calculate an unknown range, R_2, from a known range, R_1:

$$\frac{R_2}{R_1} = \frac{\rho_1}{\rho_2} \frac{\sqrt{A_2}}{\sqrt{A_1}} \qquad (58)$$

The effective atomic weight of a chemical compound or mixture can be calculated using

$$\sqrt{} \qquad (59)$$

where f_1, f_2, etc., are atomic fractions of the elements in the material. For example, for H_2O, $\sqrt{A} = 2.0$.

A tabulation of mass stopping powers ($1/\varrho$ dE/dx (MeV cm^2/g)) and ranges of protons in several materials are given in Attix and Roesch[5]; more complete data sources are listed in Holt.[12]

IV. INTERACTION OF ELECTRONS WITH MATTER

Electrons and positrons moving through a material experience nearly identical energy loss so that the term "electron" in the discussion to follow will refer to positrons as well. At rest the positron combines with an electron (annihilation), converting the rest mass of each into electromagnetic radiation.

Electrons, like heavy charged particles, lose a substantial portion of their energy through collisions with atomic electrons in the material. In fact the discussion of stopping power (energy loss per unit distance traveled) for heavy charged particles applies to electrons with few modifications. The energy loss rate of low energy electrons (less than 200 keV) is equal to that of protons of the same speed.

In general, electron energy loss is more complex than the excitation-ionization process of heavier particles. Since electron collisions are between particles of equal mass, the energy can change by a large fraction in a single collision and paths through an absorber show large deflections.

Comparison between electron and heavy particle stopping power can be made in two energy regions: (1) nonrelativistic, $\beta = v/c \ll 1$, and $T \ll m_o c^2$ ($m_o c^2 = 511$ keV for electrons and 938 MeV for protons), and (2) relativistic $\beta \approx 1$, T larger than rest energy.

In the nonrelativistic region ($\beta \approx 0$), stopping power for electrons ($z = 1$) becomes:

$$- \frac{dE}{dx} = \frac{4\pi e^4 NZ}{m_o v^2} \ell n \left(\frac{m_o v^2}{2\bar{I}} \sqrt{\frac{e}{2}} \right) \tag{60}$$

where $e = 2.718$, the base of natural logarithms. Modification of the original formula (Equation 52) is necessary since the incident and struck particles are now the same so that separation of the two after a collision is uncertain.[14] Inclusion of the correction term $\sqrt{e/2}$ has a slight effect on numerical values of the stopping power because it appears in the ℓn portion of the formula ($\ell \sqrt{e/2} = 0.15$) and the $\frac{1}{2}m_o v^2$ kinetic energy term should be much larger than the mean ionization potential, $\bar{I}$.

At highly relativistic energies ($\beta \approx 1$, and $v \approx c$), the equations become:

$$- \frac{dE}{dx} = \frac{2\pi e^4 NZ}{m_o c^2} \left\{ 2\ell n \left(\frac{2m_o c^2}{\bar{I}} \right) - 4\ell n \sqrt{1 - \beta^2} - 2 \right\} \tag{61}$$

for protons and

$$- \frac{dE}{dx} = \frac{2\pi e^4 NZ}{m_o c^2} \left\{ 2\ell n \left(\frac{2m_o c^2}{\bar{I}} \right) - 3\ell n \sqrt{1 - \beta^2} - \ell n 8 + \frac{1}{8} \right\} \tag{62}$$

for electrons. The dominant term of the bracket {} in each case is ($2m_o c^2/\bar{I}$), so in this energy region, also, the stopping power of electrons and protons is quite similar.

In addition to energy loss by electron-electron collisions, a second process, energy loss by radiation becomes important at higher energies. A charged particle will emit

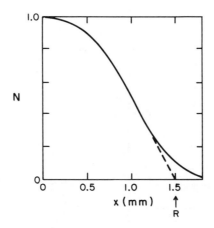

FIGURE 15.   Number-distance curve for
1 MeV electrons in aluminum.

electromagnetic radiation (Bremsstrahlung) when accelerated. Radiation loss by heavy particles is generally insignificant since accelerations in collisions with atomic electrons are small. In comparison, electrons will radiate an appreciable fraction of their energy when accelerated in the strong electric field of a nucleus. Radiation loss is not appreciable compared to collision loss except at high electron energies, the ratio being given approximately by

$$\frac{(dE/dx)_{rad}}{(dE/dx)_{col}} = \frac{ZT}{600} \qquad (63)$$

where T is the kinetic energy in MeV of the electron moving through a medium of atomic number Z.

The number-distance curve for 1 MeV electrons in aluminum is given in Figure 15. The curve has a form between the flat shape for heavy particles (Figure 12) and an exponential decrease of gamma-rays (Equation 34). This is a reasonable consequence of the fact that energy loss per collision by electrons can vary between a small fraction of the total (as for heavy particles) to all of the initial energy (as for γ-rays). The path of an electron is very irregular with many changes in direction due to large angle scattering and the total distance covered in coming to rest is much larger than its penetration distance (range) into the absorber. Electrons emitted in radioactive decays (β-rays) are not monoenergetic but are distributed in energy up to a limit characteristic of the particular nuclide. Thus the number-distance curve for β-rays is different than that of Figure 15; range in this case refers to the penetration distance for the most energetic of the β-rays. Useful empirical formulas[14] which relate maximum electron energy, E (MeV) and range in aluminum, R (g/cm^{-2}) are

$$R = 0.407\, E^{1.38} \qquad 0.15 \leqslant E < 0.8\,MeV \qquad (64)$$

$$R = 0.542\, E - 0.133 \qquad 0.8 \leqslant E < 3\,MeV \qquad (65)$$

Range-energy curves for electrons in nitrogen and aluminum are given in Figure 16.

Tables of energy loss and range of electrons and positrons can be found in the following: Nelms,[16] Berger and Seltzer,[17] with additional references in Holt.[12]

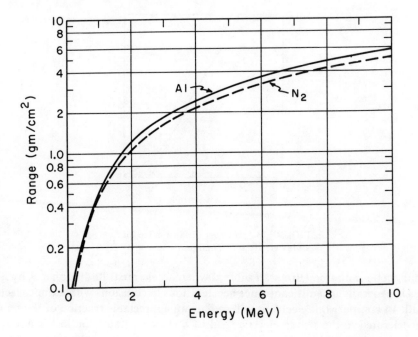

FIGURE 16.   Range-energy for electrons in nitrogen and aluminum; range-energy values for oxygen differ by less than 1% from those of nitrogen, so the curve is also valid for air (density at 20°C and 760 mm is 1.21×10^{-3} g/cm³); density of aluminum is 2.7 g/cm³.

V. INTERACTION OF NEUTRONS WITH MATTER

Neutrons and protons as the constituents of atomic nuclei form a group of particles called "nucleons". The neutron which has zero net electric charge has a slightly greater mass than the proton (1.675×10^{-24} g and 1.673×10^{-24} g, respectively). Stable within the nucleus, a neutron is unstable in free space and decays to produce a negative beta particle and a proton. The process has a half-life of 12.8 sec.

In contrast with energy loss of charged particles by excitation and ionization of atomic electrons, neutrons being electrically neutral and subject only to the strong nuclear force must interact with nuclei (or more precisely, with nucleons that make up the nuclei). The nuclear force, although of greater strength than electromagnetic forces, acts over a shorter distance which means the neutron must "collide" with the nucleus or penetrate the nucleus to interact. The four main types of neutron interactions are discussed in the following:

1.   Elastic Scattering — the predominant energy-loss process for neutrons of 1 MeV or less is a "billiard ball" collision in which energy is shared between the scattered neutron and recoil atom. The amount of energy transferred to the atom in a given collision varies with the recoil directions and is maximum for a "head-on" collision in which the atoms proceeds in the direction of the incident neutron which rebounds in the opposite direction. The fractional energy loss for the neutron is greatest for collisions with hydrogen and decreases as the mass of the moderating material is increased. The fractional energy loss for a head-on collision is

$$\left(\frac{\Delta T}{T_0}\right) = \frac{4A}{(A+1)^2} \qquad (66)$$

where T_o is the initial kinetic energy and A is the atomic mass number of the struck atom. The neutron energy decreases exponentially with the number of collisions according to the relationship:[4]

$$T_n = T_o \, e^{-n\zeta} \tag{67}$$

where T_n is the average neutron energy after n collisions and ζ is called the "logarithmic decrement."

$$\zeta = 1 - \frac{(A-1)^2}{2A} \, \ell n \left(\frac{A+1}{A-1} \right) \tag{68}$$

The expression is invalid for hydrogen, A = 1, for which ζ = 1; for heavy absorbing materials, A >> 1, the formula becomes ζ = 2/A. The number of collisions required to reduce a 1 MeV neutron to thermal energy (1/40 eV) is 18 in hydrogen, 110 in carbon, and 1800 in lead. Scattering cross sections tend to increase as the neutron energy decreases. For example, σ = 5 × 10^{-24} cm^2 for 1 MeV neutrons on hydrogen which increases to 80 × 10^{-24} cm^2 at 1/40 eV. Thus the average distance between collisions decreases as the neutron loses its energy.

2. Inelastic Scattering — in addition to recoil energy, it is possible for the nucleus to absorb energy in a collision as internal excitation which is later emitted as gamma-rays. The process is analogous to inelastic atomic scattering (excitation) where bound electrons are forced into higher orbits and subsequently return to the stable configuration by emitting X-rays. Inelastic scattering becomes competitive with elastic scattering at neutron energies above a few MeV.

3. Nuclear Reactions — a neutron can penetrate the nucleus and a proton, alpha particle or another neutron can be ejected. These processes are designated as (n,p), (n,α), and (n,2n) nuclear reactions, respectively, and in each case the original absorbing nucleus is converted to a new nuclear species which is radioactive. Although (n,p) and (n,α) reactions are possible at thermal energy in a few elements, (e.g., ^{4}He, ^{6}Li, and ^{10}B) neutrons of several MeV energy are generally necessary to have nuclear reactions occur.

4. Neutron Capture — the most common fate of low energy (eV) neutrons is to be absorbed by nuclei. The neutron becomes bound to the nucleus and the binding energy is emitted in the form of gamma-rays. This process is designated as a (n,γ) reaction, radiative capture. Since the cross section for capture varies inversely with the neutron velocity, the process becomes important at thermal energies. Capture in hydrogen results in the formation of stable deuterium and the emission of a 2.2 MeV gamma-ray.

A typical sequence of events for a neutron of initial energy of several MeV is for the neutron to dissipate its energy by elastic collisions until thermal energy is attained. This process takes 7 × 10^{-6} sec for a 2 MeV neutron in water during which time the neutron has traveled a distance of roughly 6 cm. Once at thermal energy, additional scatterings occur but do not result in appreciable energy loss. Finally, the neutron disappears by nuclear capture. The time elapse between arrival at thermal energy and capture in water is 2.4 × 10^{-4} sec, considerably longer than the slowing-down time. The neutron travels a total distance of 50 cm between thermalization and capture.

In hydrogenous materials, hydrogen plays the dominant role in the slowing-down and capture process. As an example: for neutrons at thermal energy, hydrogen scattering and capture cross sections are 80 × 10^{-24} cm^2 and 0.32 × 10^{-24} cm^2, respectively;

these cross sections for oxygen are considerably smaller, 4×10^{-24} cm^2 and 0.9×10^{-27} cm^2.

A detailed discussion of neutron physics is in Part VII of Segre,[19] studies directed toward medical applications and dosimetry are in Attix and Roesch[5] and Rossi.[18] Tables of neutron cross sections can be found in Section 8h of Harvey and Hughes[20] and additional references are in Holt.[12]

REFERENCES

1. Born, M., *Atomic Physics,* Hafner, Hew York, 1959.
2. Johns, H. E. and Laughlin, J. S., Interaction of radiation with matter, in *Radiation Dosimetry,* Hine, G. J. and Brownell, G. L., Eds., Academic Press, New York, 1956, chap. 2.
3. Heitler, W., *The Quantum Theory of Radiation,* Oxford University Press, London, 1944.
4. Segre, E., *Nuclei and Particles,* 2nd ed., W. A. Benjamin, Reading, Mass., 1977, chap. 2.
5. Attix, G. H. and Roesch, W. C., Eds., *Radiation Dosimetry,* Vol. 1, Academic Press, New York, 1968.
6. Evans, R. D., *The Atomic Nucleus,* McGraw-Hill, New York, 1955.
7. Hubbell, J. H., Photon cross sections, attenuation coefficients and energy absorption coefficients from 10 keV to 100 GeV, *Natl. Stand. Ref. Data Ser. Natl. Bur. Stand.,* 29, August 1969.
8. Hubbell, J. H., Photon mass attenuation and mass-energy absorption coefficients for H, C, N, O, Ar, and seven mixtures from 0.1 keV to 20 MeV, *Radiat. Res.,* 70, 58, 1977.
9. Evans, R. D., Gamma rays, in *American Institute of Physics Handbook,* 2nd ed., Gray, D. E., Ed., McGraw-Hill, New York, 1963, Sect. 8f.
10. Marmier, P. and Sheldon, E., *Physics of Nuclei and Particles,* Vol. I, Academic Press, New York, 1969.
11. Grodstein, G. W., X-Ray attenuation coefficients from 10 keV to 100 meV, *Natl. Bur. Stand. (U.S.) Circ.,* 583, 1957.
12. Holt, D. D., Ed., Sources of atomic and nuclear data for biomedical purposes, *Phys. Med. Biol.,* 24, 1, 1979.
13. Bethe, H. A., *Z. Physik,* 76, 293, 1932.
14. Segre, E., Ed., *Experimental Nuclear Physics,* Vol. I, John Wiley & Sons, New York, 1953.
15. Bethe, H. A. and Ashkin, J., Passage of radiation through matter, in *Experimental Nuclear Physics,* Vol. I (Part II), Segre, E., Ed., John Wiley & Sons, New York, 1953.
16. Nelms, A. T., Energy loss and range of electrons and positrons, *Natl. Bur. Stand. Circ.,* 577 (Suppl.), July 30, 1958.
17. Berger, M. J. and Seltzer, S. M., Tables of energy losses and ranges of electrons and positrons, *NASA Spec. Publ.,* SP-3012, 1964.
18. Rossi, H. H., Neutrons and mixed radiations, in *Radiation Dosimetry,* Hine, G. J. and Brownell, G. C., Eds., Academic Press, New York, 1956.
19. Segre, E., Ed., *Experimental Nuclear Physics,* Vol. II, John Wiley & Sons, New York, 1953.
20. Harvey, J. A. and Hughes, D. J., Neutrons, in *American Institute of Physics Handbook,* Gray, D. E., Ed., McGraw-Hill, New York, 1963.
21. Weber, J. and Van den Berge, D. J., The effective atomic number and the calculation of the composition of phantom materials, *Br. J. Radiol.,* 42, 378, 1969.

Chapter 4

MATHEMATICS OF RADIOACTIVE DECAY

B. T. A. McKee

TABLE OF CONTENTS

I. Continuum Theory of Radioactive Decay.............................107
    A. Continuum Theory — One Species 107
        1. Exponential Decay and the Decay Constant λ 107
        2. The Half-Life T $_{1/2}$ and the Mean Life τ 108
        3. Partial Decay Constants 109
        4. Units of Radioactivity..................................110
        5. Examples...110
            a. Graphical Representation 110
            b. Example Involving Simple Decay....................110
    B. Continuum Theory — More Than One Species 112
        1. Growth of a Daughter 112
        2. Ratio of Activities of Parent and Daughter 114
            a. Daughter Longer-Lived Than Parent ($\tau_2 > \tau_1$ or $\lambda_2 <$
                λ_1)..114
            b. Daughter and Parent with Nearly Equal Decay
                Constants 115
            c. Daughter Shorter-Lived Than Parent ($\tau_2 < \tau_1$ or $\lambda_2 >$
                λ_1)..115
            d. Daughter Much Shorter-Lived Than Parent ($\tau_2 \ll \tau_1$ or λ_2
                $\gg \lambda_1$)..116
        3. Production of Radionuclides by Nuclear Reactions 117
        4. General Equations of Radioactive Growth and Decay 119
        5. Accumulation of Stable End Products 121
        6. Graphical Solutions...................................122
        7. Numerical Examples 122
            a. ^{14}C Production and Decay 122
            b. ^{131}I Decay 124
            c. Impurities in ^{123}I...............................125
            d. The ^{99}Mo $- ^{99m}$Tc Generator........................125

II. Statistical Fluctuations in Radioactive Decay.........................126
    A. Frequency Distributions 127
        1. Location Indexes 127
        2. Dispersion Indexes.....................................128
    B. Mathematical Models of Frequency Distributions................132
        1. The Binomial Distribution 132
        2. The Poisson Distribution 135
        3. The Normal Distribution 139
        4. The Interval Distribution 143
        5. Summary of Frequency Distributions 144
        6. Some Numerical Examples 144
            a. A Counting Experiment 145
            b. A Counting-Loss Problem 146

              c.   Random-Coincidence Rate.........................147
    C.   Some Further Statistical Considerations........................148
         1.   Propagation of Errors..................................148
         2.   Parameter Fitting — Method of Maximum Likelihood 150
         3.   Tests for Goodness of Fit 152
              a.   The t-Test for Consistency of Means................152
              b.   The F-Test for Consistency of Standard Deviations ...152
              c.   The Chi-Square (χ^2) Test for Consistency of a Model
                     Distribution 153
         4.   Some Numerical Examples 154
              a.   The Contribution of Background in a Counting
                     Measurement 154
              b.   Consistency of Two Means.......................155
              c.   A Chi-Square Test 155
              d.   Determination of a Decay Constant by Least Squares and
                     Maximum Likelihood Methods 156

References...159

There is a quotation by E. T. Wittaker, "Everybody believes in the exponential law of errors; the experimenters because they think it can be proved by mathematics; and the mathematicians because they believe it has been established by observations." The same might well be said regarding radioactive decay, which is, of course, intimately related to probability theory.

This chapter is intended to provide for the experimenter some knowledge of the mathematical relationships involved in nuclear decay and in the measurements of radioactive processes. We can treat the laws of radioactive decay independent of the emission accompanying the decay and independent of the physical details of the transformation itself. This is possible because the law of decay is independent (to an exceedingly good approximation) of the mechanism of the transformation. The results thus acquired are widely applicable.

The random nature of radioactive decay is by no means unique to this area; indeed, much of our physical world involves at a microscopic or submicroscopic level, processes of a random nature. However, the processes of radioactive decay involve the release of relatively large energies and thus it is easy to observe single random events, and to appreciate the need for an analysis based on statistical theory.

Despite the fundamentally random and discrete nature of radioactive decay, general mathematical relations may be derived which model the decay and growth of radioactive species in terms of continuous variables. This is done by neglecting the fact that every substance contains a discrete number of atoms and by treating this number as a continuous variable. This procedure, which will be followed in Section I of this chapter is a good approximation for processes involving a great number of atoms. Furthermore, if we consider many ensembles, initially identical, of radioactive atoms, then the continuum theory yields exactly the average number of atoms contained in the various ensembles at any subsequent time. However, each system will depart somewhat from the average and for the elucidation of these statistical fluctuations it is necessary to take into account the discontinuous and discrete nature of the radioactive decay, which will be done in Section II.

I. CONTINUUM THEORY OF RADIOACTIVE DECAY

A. Continuum Theory — One Species

1. Exponential Decay and the Decay Constant λ

The exponential decay of a radioactive species was first observed at the turn of the century by Rutherford and Soddy at McGill. They found that with an air current a radioactive gas (now known as ^{220}Rn) could be swept away from thorium oxide into an ionization chamber. When they stopped the air current, the conductivity of the chamber diminished with time according to a geometrical law, falling to half value in about 1 min. Further investigations through the years on other radioactive species have shown that the decay of a given mass of material is accurately exponential. This is true for the disappearance by decay of short-lived unstable particles and excited states, as well as of long-lived radioactive elements. A result of such general application must surely have some general basis, and indeed, the decay law of radioactive substances which leads to the observed exponential behavior was formulated by Rutherford and Soddy[1] on the basis of their early studies on the thorium decay series.

The fundamental law of radioactive decay can be formulated thus: given a radioactive atom, the probability that it will decay during the time interval dt is λdt. The constant λ is called the *decay constant*. It has the dimensions of reciprocal time, and it is characteristic of the given nuclear species and of the mode of decay. The constant is independent of the age of the atom and, being a nuclear property, is not affected* by chemical or physical characteristics such as changes in temperature or pressure. This type of law is characteristic of *casual* events and applies to all types of radioactive decay — alpha, beta, gamma, electron capture, and spontaneous fission.

To relate this fundamental law to the behavior of a radioactive species, we can consider a single radioactive substance that has, initially, a large number N(0) atoms. As N(0) is assumed a large number, we may consider N(t), the number of atoms at time t, to be a continuously variable quantity. Then, according to our fundamental law, the decrease in the number of atoms present, $-dN(t)$, due to their decay during the time interval dt, is N(t)λ dt. This can be written

$$dN(t) = -N(t) \lambda dt \tag{1}$$

or in integral form,

$$\int_{t=0}^{t} \frac{dN(t)}{N(t)} = -\int_{t=0}^{t} \lambda dt \tag{2}$$

Recalling that λ is assumed constant with time, we can integrate using the boundary condition that N(t) equals N(0) at time t = 0 and obtain

$$\ln \frac{N(t)}{N(0)} = -\lambda t \tag{3}$$

or, in the usual exponential form,

$$N(t) = N(0) e^{-\lambda t} \tag{4}$$

---

* There are minor exceptions to this statement in cases of decay through internal conversion of electrons and through capture of K-electrons. Because they involve the charge density of orbital electrons overlapping the nucleus, they are very slightly susceptible to the chemical state of the atom.[2]

This Equation 4 can be considered the mathematical formulation of the fundamental law of radioactive decay.

A derivation based on the laws of chance is also possible. Since λdt is the probability that a particular atom will decay in the time dt, then its chance of survival for that time is $(1 - \lambda dt)$, for a time 2dt is $(1 - \lambda dt)^2$, ... , and for any arbitrary time t = ndt is $(1 - \lambda dt)^n = (1 - \lambda dt)^{t/dt}$. (The individual probabilities multiply for independent events.) If the arbitrary time interval dt is taken as very small, then a series expansion yields

$$\underset{dt/t \to 0}{\text{limit}} \ (1 - \lambda dt)^{t/dt} = e^{-\lambda t} \tag{5}$$

Thus the probability of survival during the time t is $e^{-\lambda t}$ for each atom. In a group of atoms we would expect statistical fluctuations in the actual fraction which survives the time t, such that the average fraction which survives, $N(t)/N(0)$, is also $e^{-\lambda t}$.

Two assumptions have been implicit in these derivations of the exponential decay law:

1.   The decay constant λ is the same for all atoms of the species.
2.   The decay constant λ is independent of the age of the particular atom.

These two conditions have been shown to be both necessary[3] and sufficient[4] for the derivation of the exponential decay law. The strongest evidence for the validity of these assumptions comes from the large body of experimental data which verifies the exponential nature of radioactive decay. However, there is a theoretical prediction[5] that the decay constant in alpha decay is no longer a constant after a large number of mean lives have elapsed, but after such a length of time there would be not enough atoms to permit observation of the deviation from exponential decay.

The decay constants for radioactive nuclides extend between 1.6×10^{-18} sec^{-1} and 3×10^6 sec^{-1}, a range of over 10^{24}. If excited nuclear states (which decay by the same exponential law) are included then the range of λ increases upwards to 10^{16} sec^{-1}.

We have seen that the number N of a particular species of radioactive atoms decays exponentially with time and with a characteristic decay constant λ. Usually we are more interested in the number of disintegrations per unit time, called the *activity*, because our measuring apparatus, be it scintillation counter, geiger counter, or dosimeter responds in proportion to the intensity of radiation or activity.

Of course the activity, which we will denote I(t), is simply the rate of decrease in the number of radioactive atoms, $-dN(t)/dt$, which is equal to $\lambda N(t)$. If we denote the initial activity by I(0), then it follows directly from Equation 4 that

$$I(t) = I(0) \, e^{-\lambda t} \tag{6}$$

so that, as one would intuitively expect, the activity follows the same exponential decay behavior as the number of radioactive atoms.

2. The Half-Life $T_{1/2}$ and the Mean Life τ

The *half-life* $T_{1/2}$ (or, more properly, *half-period*) is the time interval over which the chance of survival of a particular radioactive atom is exactly one half. Then, if λ is the decay constant, Equation 3 yields

$$T_{1/2} = \ln 2/\lambda = 0.693/\lambda \tag{7}$$

For a large initial number N(0) of radioactive atoms, with initial activity N(0)λ, the average value of the activity one half-life later, N(T$_{1/2}$)λ, is N(0)λ/2, or one half the initial activity.

The actual life of any particular radioactive atom may range between zero and infinity. The *average lifetime* of a large number of similar atoms is, however, a definite and useful quantity. For N(0) radioactive atoms present at time zero, we have seen that the number remaining undecayed at the subsequent time t is N(t) = N(0)e$^{-\lambda t}$. All of these remaining have a lifetime longer than t. Those which decay within the short time interval dt following t can be considered to have a lifetime t, and these will be of number N(t)λdt = N(0)λe$^{-\lambda t}$. The total lifetime of all the atoms is obtained by integrating the product of the lifetime t with the number having this lifetime over all values of t from 0 to ∞, and the average lifetime, which is called the *mean life* τ, is simply this total lifetime divided by the initial number N(0).

$$\tau = \frac{1}{N(0)} \int_0^\infty tN(0)\lambda e^{-\lambda t}dt = 1/\lambda \qquad (8)$$

Thus the mean life τ is simply the reciprocal of the decay constant λ. The mean life can be related to the half-life through Equation 5.

$$\tau = T_{1/2}/0.693 = 1.44\,T_{1/2} \qquad (9)$$

Substitution in the exponential decay Equation 4 shows that in one mean lifetime the number of radioactive atoms, or their activity, falls to e^{-1} = 0.368 of the initial value.

The total number of radioactive atoms present at a particular time N(t) is simply the product of the total activity I(t) and the mean life τ, because

$$I(t)\tau = \frac{N(t)\lambda}{\lambda} = N(t) \qquad (10)$$

3. Partial Decay Constants

Many radioactive species can decay via more than one route. For example, ^{64}Cu can decay by electron emission (β^- decay), positron emission (β^+ decay), or electron capture. If the competing modes of decay of any nuclide have probabilities λ_1, λ_2, λ_3... per unit time, then the total probability of decay is represented by the total decay constant λ, where

$$\lambda = \lambda_1 + \lambda_2 + \lambda_3 + \ldots \qquad (11)$$

The *partial activity* of a sample of N(0) nuclei, if measured by a method susceptible to one particular mode of decay characterized by λ_i, is

$$I_i(t) = -\frac{d\,N_i(t)}{dt} = \lambda_i\,N(t) = \lambda_i\,N(0)\,e^{-\lambda t} = I_i(0)\,e^{-\lambda t} \qquad (12)$$

and the total activity as a function of time is

$$I(t) = -\frac{d\,N(t)}{dt} = \sum_i \frac{d\,N_i(t)}{dt} = N(t) \sum_i \lambda_i$$

$$= \lambda N(0)\,e^{-\lambda t} = I(0)\,e^{-\lambda t} \qquad (13)$$

It should be noted from Equation 13 that the partial activities, such as positron emission from ^{64}Cu, are proportional to the total activity at all times. Each partial activity decays with time as $e^{-\lambda t}$, not as $e^{-\lambda_i t}$. This is because the decrease of activity with time is due to the depletion of the stock of atoms N(t), and this depletion is accomplished by the combined action of all the competing modes of decay.

4. Units of Radioactivity

The *curie* (Ci) was originally defined with reference to the activity of 1 g of radium. This was, of course, susceptible to periodic revision so that the curie was redefined in 1950 as "the quantity of any radioactive nuclide in which the number of disintegrations per second is 3.700×10^{10}."

Note that the number of disintegrations referred to in the definition involves the sum of all competing modes of disintegration. Therefore the full decay scheme of a nuclide must be known before the number per second of a particular decay mode such as β or γ rays can be predicted from a specified activity of radionuclide, or before the quantity of a radioactive sample can be expressed in curies as a result of measurements on one particular decay mode.

Note also that many radioactive nuclei disintegrate in a cascade involving the emission of multiple rays. These must be all considered when, for example, calculating the dose rate from a given amount of activity.

5. Examples

a. Graphical Representation

Some of the relations outlined in this section are illustrated in the two graphs of exponential decay, Figures 1 and 2. The first is a plot of activity (such as one might measure with a hypothetical scintillation counter of 100% efficiency) vs. time for any radioactive species. It is noted that the total area under this decay curve of activity vs. time equals the total number of radioactive atoms initially present:

$$\text{Area} = \int_0^\infty I(t)dt = \int_0^\infty \lambda N(0) e^{-\lambda t} \, dt = N(0) \qquad (14)$$

This area is also the same as within the rectangle $I(0)\tau$, so that if the initial activity $I(0)$ were to remain constant, all of the radioactive atoms would be depleted after one mean lifetime τ.

If the logarithm of $I(t)$ (or $N(t)$) is plotted vs. time, then a straight line is obtained with slope $-\lambda$:

$$\ell n \, I(t) = \ell n \, N(0) - \lambda t \qquad (15)$$

This can be conveniently and directly performed using semilogarithmic graph paper. An illustration is shown in Figure 2.

b. Example Involving Simple Decay

Assume that in vitro scintillation counting measurements are being performed using two different radionuclides as labels. Each decays simply via emission of a single gamma ray. The two hypothetical radionuclides A and B have half-lives of 300 days and 2 hr, respectively. The detection efficiency of the scintillation counter is 10% for the gamma rays from isotope A and 5% for the higher energy gamma rays from B. If we want to obtain one million counts from each radionuclide and are willing to count overnight (10 hr), we can predict the amount of labeled activity required for each of these two nuclides. Equation 6 expressed the activity $I(t)$ in terms of the initial activity $I(0)$. In the case of A, the half-life is much greater than our measuring time, so that

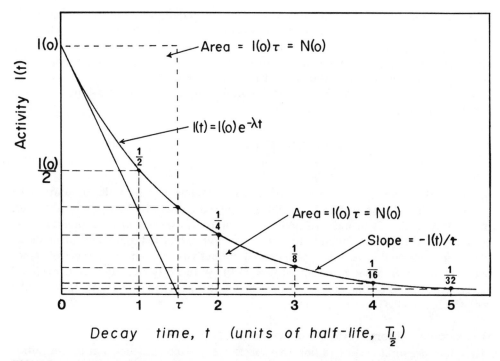

FIGURE 1.   Exponential decay of the activity from a typical radionuclide. Some relationships involving the initial activity, total activity, and mean life are indicated.

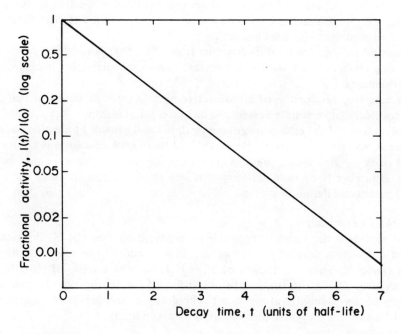

FIGURE 2.   A semilogarithmic representation of radioactive decay. The slope of the line, log $[I(t)/I(0)]/t$, equals $-\lambda$ log e, or in natural logarithms, ln $[I(t)/I(0)]/t = -\lambda$.

to a very good approximation we can consider the activity as constant during the 10 hr of counting. Thus, to obtain one million counts in 10 hr having 10% efficiency in the counter requires one million gamma rays per hour. Because the decay scheme is simple, this is equivalent to one million disintegrations per hour, or 278 disintegrations per second, which amounts to 0.0075 μCi. In the case of B, the activity is decreasing rapidly during the measurement, so that we should integrate the expression for activity I(t), (Equation 6) over the 10 hr counting time.

$$I_{measured} = \text{Efficiency} \times \int_{t=0}^{10} I(0)\, e^{-\lambda t}\, dt \qquad (16)$$

where we can leave the time in units of hours or change it to seconds. In hours, $\lambda(B) = 1/\tau = 1/(1.44T_{1/2}) = 0.35$ hr^{-1}, and the integral reduces to $2.88\,(-e^{-3.5} + 1)\, I(0)$, or $2.8\, I(0)$, with $I(0)$ in units of activity per hour. As we would have predicted, by counting this short-lived nuclide for 10 hr we have exhausted almost all of its decay (97%) and we could have well approximated this activity by the total number of radioactive atoms of B initially present $N(0) = \tau I(0)$. With a 5% counting efficiency, and $I_{measured}$ required to be one million events, then in this case $I(0) = 7.1 \times 10^6$/hr or 2000/sec, which is 0.05 μCi of initial activity.

B. Continuum Theory — More Than One Species

In a number of cases one radioactive species decays into another that is also radioactive. The first is called the parent, or mother, species and the second is called the daughter. In turn, the daughter's decay product may be radioactive so that the family continues for several generations. Sometimes one radioactive species can decay by two processes, e.g., by alpha and beta emission, giving rise to two different daughter substances: this occurrence is called branching.

The natural radioactive families (starting from ^{238}U, ^{235}U, and ^{232}Th) provide examples of long chains of radioactive decays, as do also the successive decays of typical fission fragments.

The accelerator production of a radioactive nuclide through a nuclear reaction can also be considered an example of a parent-daughter relationship.

A general form of the equations governing decay and growth of members of a radioactive series was first outlined by Bateman[6] and these general solutions are frequently referred to as the Bateman equations. Before presenting the general equations applicable to a number N of radioactive species, we shall consider in detail the relations between parent and daughter species.

1. Growth of a Daughter

Let N_1 represent the number of atoms of a parent species and N_2 the number of atoms of a daughter species present at any time t, and let the corresponding decay rates be λ_1 and λ_2. Then the activity of N_1 is $N_1\lambda_1$ and the activity of N_2 is $N_2\lambda_2$. The rate of change, dN_2/dt, in the number of atoms of the second type is then equal to the supply of new daughter atoms due to the decay of the parent N_1, diminished by the rate of loss of the daughter species through its own decay:

$$\frac{dN_2}{dt} = N_1\, \lambda_1 - N_2\lambda_2 \qquad (17)$$

If the only source of the parent species is from an initial supply $N_1(0)$ at $t = 0$, then $N_1 = N_1(0)e^{-\lambda t}$ and, with these initial conditions on N_1, Equation 17 becomes

$$\frac{dN_2}{dt} = N_1(0) e^{-\lambda_1 t} - N_2 \lambda_2 \qquad (18)$$

The general solution of this differential equation will be of the form

$$N_2(t) = N_1(0) (A_1 e^{-\lambda_1 t} + A_2 e^{-\lambda_2 t}) \qquad (19)$$

Substituting from Equation 19 into Equation 18 yields

$$A_1 = \frac{\lambda_1}{\lambda_2 - \lambda_1} \qquad (20)$$

The coefficient A_2 depends on the value of N_2 at $t = 0$. For the important special case in which $N_2 = 0$ at $t = 0$, we have from Equation 19

$$A_1 + A_2 = 0 \qquad (21)$$

Hence for this case $A_2 = -A_1$ and we have

$$N_2(t) = N_1(0) \frac{\lambda_1}{\lambda_2 - \lambda_1} (e^{-\lambda_1 t} - e^{-\lambda_2 t}) \qquad (22)$$

as the solution $N_2(t)$ for initial conditions $N_1 = N_1(0)$ and $N_2 = 0$ at $t = 0$.

The decay activity I_2 of the daughter is $N_2 \lambda_2$ (not $-dN_2/dt$) so that

$$I_2(t) = N_1(0) \frac{\lambda_1 \lambda_2}{\lambda_2 - \lambda_1} (e^{-\lambda_1 t} - e^{-\lambda_2 t}) \qquad (23)$$

or, since the activity $I_1(t)$ of the parent is $N_1 \lambda_1 = N_1(0) \lambda_1 e^{-\lambda_1 t}$,

$$I_2(t) = I_1(t) \frac{\lambda_2}{\lambda_2 - \lambda_1} (1 - e^{-(\lambda_2 - \lambda_1)t}) \qquad (24)$$

It may be informative to derive the equation of growth for a daughter species (Equation 22) by an alternative route. At time $t = 0$, let $N_1 = N(0)$ and $N_2 = 0$. Then at a later time, $t = t'$, there will remain $N_1(t') = N_1(0) e^{-\lambda_1 t'}$ atoms of the parent species. In the time interval between t' and $t' + dt'$, the number of new daughter atoms formed will be $N_1(t') \lambda_1 dt'$. The fraction of these daughter atoms which survive until a later time t is $e^{-\lambda_2(t-t')}$. Then the total stock of the daughter species at time t is given by an integral over all values of the time t' between $t = 0$ and $t = t$, and is

$$N_2(t) = \int_0^t (N_1(t') \lambda_1 \, dt') (e^{-\lambda_2(t-t')}) \qquad (25)$$

Thus

$$N_2(t) = N_1(0)\lambda_1 e^{-\lambda_2 t} \int_0^t e^{-(\lambda_1 - \lambda_2)t'} dt' = N_1(0) \frac{\lambda_1}{\lambda_2 - \lambda_1} (e^{-\lambda_1 t} - e^{-\lambda_2 t}) \qquad (26)$$

in agreement with Equation 22.

The amount of daughter product $N_2(t)$ is evidently zero both at times $t = 0$ and at $t = \infty$ when all of the atoms of both parent and daughter have decayed. At some intermediate time t_m the amount of the daughter $N_2(t)$ and hence its activity $I_2(t) = N_2(t) \lambda_2$, passes through a maximum. At this time t_m the derivative of N_2 will be zero, $dN_2/dt = 0$. From Equation 22 we obtain for this condition

$$\lambda_1 e^{-\lambda_1 tm} = \lambda_2 e^{-\lambda_2 tm} \qquad (27)$$

from which it follows that the time t_m of maximum activity of the daughter product is

$$t_m = \frac{\ln(\lambda_2/\lambda_1)}{\lambda_2 - \lambda_1} \qquad (28)$$

assuming that $N_1 = N_1(0)$ and $N_2 = 0$ at $t = 0$.

It is evident that t_m is positive and real for either $\lambda_1 > \lambda_2$ or $\lambda_1 < \lambda_2$. If λ_1 and λ_2 are approximately equal ($\lambda_2 = \lambda_1 + \Delta$, where $\Delta \ll \lambda_1$) then t_m can be evaluated by a series expansion

$$\ln \left(1 + \frac{\Delta}{\lambda_1}\right) = \frac{\Delta}{\lambda_1} - \frac{\Delta^2}{2\lambda_1^2} + \cdots\cdots\cdots$$

so that

$$t_m \simeq \frac{1}{\lambda_1} \left(1 - \frac{\Delta}{2\lambda_1}\right) \simeq \frac{1}{\lambda_1} \left(\frac{\lambda_1}{\lambda_2}\right)^{1/2} = \sqrt{\tau_1 \tau_2} \qquad (29)$$

In this special case where parent and daughter have similar decay constants, the maximum activity of the daughter occurs about one mean lifetime after $t = 0$.

Whatever the relative decay constants, we have at time t_m (from Equation 17)

$$N_1(t_m)\lambda_1 = N_2(t_m)\lambda_2 \quad \text{or} \quad I_1(t_m) = I_2(t_m) \qquad (30)$$

That is, the activity of the residual parent N_1 and of the accumulated daughter N_2 are equal at time t_m, and this activity is

$$I_1(t_m) = I_2(t_m) = N_1(0)\lambda_1 e^{-[\lambda_1 \ln(\lambda_2/\lambda_1)]/(\lambda_2-\lambda_1)} = N_1(0)\lambda_1 \left(\frac{\lambda_1}{\lambda_2}\right)^{\lambda_1/(\lambda_2-\lambda_1)} \qquad (31)$$

Between $t = 0$ and t_m the activity of the parent is greater than the activity of the daughter ($dN_2/dt > 0$), whereas from $t = t_m$ to $t = \infty$ the activity of the daughter exceeds that of the parent ($dN_2/dt < 0$).

2. Ratio of Activities of Parent and Daughter

Under the initial conditions of $N_1 = N_1(0)$ and $N_2 = 0$ at $t = 0$, the ratio of activities of the parent and daughter species is given directly by Equation 24. This activity ratio I_2/I_1 is zero for $t = 0$, unity for $t = t_m$, and has its maximum value for large values of t. We shall examine a few distinct cases, depending on the relative decay constants of parent and daughter.

a. Daughter Longer-Lived Than Parent ($\tau_2 > \tau_1$ or $\lambda_2 < \lambda_1$)

If $\lambda_2 < \lambda_1$, then the activity ratio I_2/I_1 increases continuously as t increases. Equation 24 may be written

$$\frac{I_2(t)}{I_1(t)} = \frac{\lambda_2}{\lambda_1 - \lambda_2} (e^{(\lambda_1-\lambda_2)t} - 1) \qquad (32)$$

In the extreme case of $\lambda_2 \ll \lambda_1$, (daughter much longer-lived than parent) then the activity of the daughter species finally becomes almost independent of the residual activity of the parent. In this limit, for $t \gg \tau_1$ Equation 32 becomes

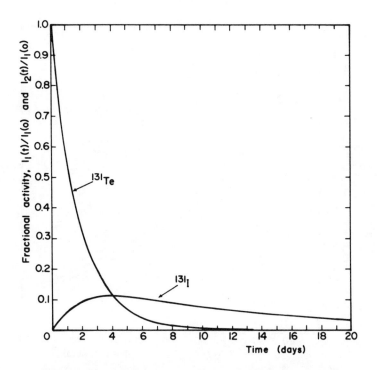

FIGURE 3.   An illustration of the growth and decay of the daughter ^{131}I (half-life 8.1 days) from its shorter-lived parent ^{131}Te (30 hr). For times greater than a few days the ^{131}I decay is almost independent of the residual parent activity. The maximum activity of the daughter occurs at about 4 days, as predicted by Equation 28, and then parent and daughter activities are equal.

$$I_2(t) \simeq N_1(0)\,\lambda_2\,e^{-\lambda_2 t} \tag{33}$$

In this approximation, the initial stock of short-lived parent atoms $N_1(0)$ becomes equivalent to an initial stock of long-lived daughter atoms $N_2(0) \simeq N_1(0)$, which decay exponentially with a rate λ_2 characteristic of the long-lived daughter. An example of the activity ratio for such a case of short-lived parent decaying to long-lived daughter is given in Figure 3.

b. Daughter and Parent with Nearly Equal Decay Constants

If λ_1 and λ_2 are nearly equal, such that $\lambda_2 = \lambda_1 + \Delta$, where $\Delta/\lambda_1 \ll 1$, then Equation 24 for the ratio of activities of daughter to parent becomes

$$\frac{I_2(t)}{I_1(t)} = \frac{\lambda_2 \Delta t}{\Delta}\left(1 - \frac{\Delta t}{2} + \ldots\ldots\right) \simeq \lambda_2 t \tag{34}$$

Thus in this case of approximately equal parent and daughter decay constants, the activity ratio would increase approximately linearly with time, so long as $t \ll 2/\Delta$.

c. Daughter Shorter-Lived Than Parent $(\tau_2 < \tau_1$ or $\lambda_2 > \lambda_1)$

If $\lambda_1 < \lambda_2$, then for all values of t which are large compared with $1/(\lambda_2 - \lambda_1)$ Equation 24 becomes

$$\frac{I_2(t)}{I_1(t)} \simeq \frac{\lambda_2}{\lambda_2 - \lambda_1} \tag{35}$$

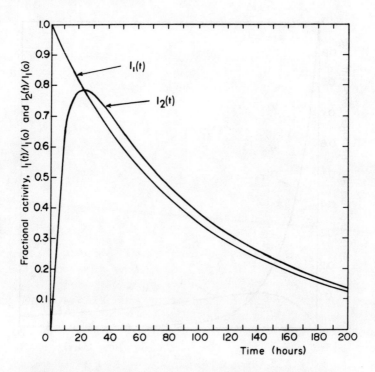

FIGURE 4.   An illustration of the growth and decay of a daughter (half-life 6 hr) from its longer-lived parent (half-life 66 hr). For times greater than 50 hr, the daughter and parent activities remain in the ratio of approximately 1.1/1 (Equation 35). This is an example of transient equilibrium; because the parent lifetime is considerably longer than the daughter's some characteristics of secular equilibrium are also illustrated. This figure gives the time relationship for the growth and decay of daughter ^{99m}Tc from parent ^{99}Mo, except in that case only 86% of the parent decays through ^{99m}Tc so that the daughter activity would need to be reduced by that factor (example in Section II.B.7.d).

When, as in this example, the ratio of activity of daughter to parent is constant, a particular type of radioactive equilibrium is said to exist. This is called *transient equilibrium* when the ratio is significantly greater than unity. This situation in which the activity of the daughter product exceeds the activity of its parent species should not be surprising. Figure 4 illustrates a typical example of transient equilibrium. The total area under each activity curve in Figure 4 from $t = 0$ to $t = \infty$ represents the total number of disintegrations for parent and daughter species. If there are no alternative decay branches for the parent, then the area under the parent and daughter activity curves must be equal as the same nuclei are involved. From $t = 0$ to $t = t_m$ the activity of the parent exceeds that of its daughter. After t_m the activity of the daughter has built up and exceeds that of its parent. The equality of the total number of disintegrations from parent and daughter can be verified by integrating Equation 23.

$$\int_0^\infty I_2(t)dt = N_1(0)\,\frac{\lambda_1\lambda_2}{\lambda_2 - \lambda_1}\,\int_0^\infty (e^{-\lambda_1 t} - e^{-\lambda_2 t})dt = N_1(0)$$

$$(36)$$

d. Daughter Much Shorter-Lived Than Parent ($\tau_2 \ll \tau_1$ or $\lambda_2 \gg \lambda_1$)

When the lifetime of the daughter is very much shorter than that of the parent ($\lambda_1 \ll \lambda_2$) then Equation 24 takes on a particularly simple form:

$$I_2(t) \simeq I_1(t) \; (1 - e^{-\lambda_2 t}) \tag{37}$$

The daughter activity $I_2(t)$ increases according to an exponential growth curve governed by its own decay constant λ_2. In this case the equilibrium ratio of activities becomes close to unity, since for $t \gg \tau_2$ Equation 37 reduces to

$$I_2(t) \simeq I_1(t) \tag{38}$$

This situation, where the parent and daughter activities reach an equilibrium in which they are approximately equal, is referred to as *secular equilibrium*. An alternative form of expressing Equation 38 is:

$$\frac{N_2}{N_1} \simeq \frac{\lambda_1}{\lambda_2} \qquad (t \gg \tau_2) \tag{39}$$

This can be interpreted as indicating that the number of daughter atoms present is approximately independent of time, for if the lifetime of the parent N_1 is sufficiently long, the number of parent atoms will not be appreciably changing.

An example of secular equilibrium and the applicability of Equation 37 can be found in uranium ores, where the lifetime of the parent ^{238}U is very long compared with the lifetime of its daughter products. Then Equation 39 applies to the daughter species and the number of atoms of a daughter is inversely proportional to its decay constant.

Common examples of secular equilibrium are provided by radionuclide generators. These involve the production of a useful short-lived daughter species from a long-lived parent that can be maintained in the laboratory.

An historically important example of secular equilibrium is provided by growth of ThX (now known as ^{224}Rn), half-life 3.64 days, from the parent RdTh (^{222}Th), half-life 1.91 years. In the experiment which was a key to their elucidation of the radioactive transformation laws,[1] Rutherford and Soddy performed a chemical separation of the ThX. They observed that the ThX activity decayed exponentially with a half-life of about 4 days, and that the activity of the thorium sample grew exponentially with the same half-life. Figure 5 is a representation of their results.

3. Production of Radionuclides by Nuclear Reactions

The production of radioactive nuclides by nuclear reactions, e.g., with an accelerator, can be considered as the growth of a daughter product from an effectively long-lived parent.

Consider the example of production of the positron-emitting nuclide ^{11}C by proton bombardment:

$$^{14}N + {}^1H \rightarrow {}^{11}C + {}^4He \tag{40}$$

or as this reaction is commonly denoted, $^{14}N(p,\alpha)^{11}C$. We shall call the number of target atoms of ^{14}N which are accessible to the proton beam N_1, and the probability of transforming one of these atoms into ^{11}C in unit time λ_1. Then $N_1\lambda_1$ is the rate at which new atoms of ^{11}C are produced. Thus the target is treated mathematically as though it were a parent source, having an activity $I_1 = N_1\lambda_1$ and producing a radioactive daughter species N_2. (The ^{11}C decays via positron emission to ^{11}B with a half-life of 20.4 min.) The probability λ_1 of producing the (p,α) reaction on ^{14}N is very small, but the number of target atoms N_1 is very large, so that the product $N_1\lambda_1$ is of intermediate magnitude. Usually a negligible fraction of the target atoms are transformed during the irradiation period so that the number of residual target atoms, $N_1(t) =$

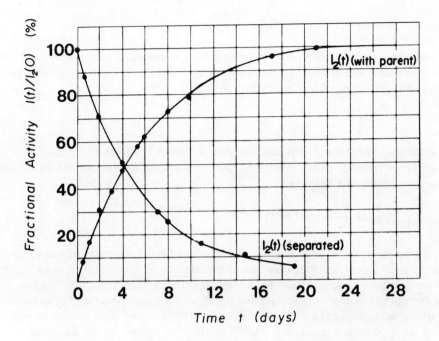

FIGURE 5.   A representation of the results of Rutherford and Soddy[1] showing the growth of the daughter ThX (^{224}Rn) with the parent RdTh (^{222}Th) following separation and showing the decay of the separated daughter. The separated daughter decays at the same rate (half-life 3.64 days) as the activity of the daughter with the parent grows, as expected from Equation 37. The parent activity (half-life 1.91 years) is almost constant on the time scale of this graph. This is an example of secular equilibrium.

$N_1(0)e^{-\lambda_1 t}$, is effectively equal to $N_1(0)$. Thus for the example above, the activity $I_2(t)$ of ^{11}C produced after a uniform bombardment of duration t will be given by Equation 37 for the growth of activity of a daughter product from a long-lived parent, although the parent activity I_1 is best replaced by the product $N_1\lambda_1$ as defined above.

$$I_2(t) \simeq N_1\lambda_1(1 - e^{-\lambda_2 t}) \tag{41}$$

The *yield* Y(t) of such a nuclear reaction is defined as the rate of production of activity of the radioactive product, under specific bombardment conditions. For example, in the ^{14}N(p,α)^{11}C reaction, the yield of ^{11}C is 7.4 mCi/min/μA of 14 MeV protons bombarding a thick target of ^{14}N.[7]

The yield is the rate at which new activity is formed. Of course as the daughter product begins to decay, its activity is depleted through its decay. Thus an expression for the yield can be obtained by differentiating the expression for the daughter activity, (such as Equation 41) and evaluating at t = 0. In the approximation of $\lambda_1 \ll \lambda_2$ where Equation 41 is valid, we have

$$Y = \frac{dI_2}{dt}\bigg|_{\text{at } t=0} = N_1\lambda_1\lambda_2 \tag{42}$$

For greater generality we can differentiate the more general expression for the activity of the daughter product given in Equation 24 and obtain the very same value $Y = N_1\lambda_1\lambda_2$. Note that the yield Y has dimensions of activity per unit time. Because the effective activity of the parent target $N_1\lambda_1$ can be expressed as Y/λ_2 or $Y\tau_2$, the net activity I_2 accumulated during a time t (Equation 41) can be written

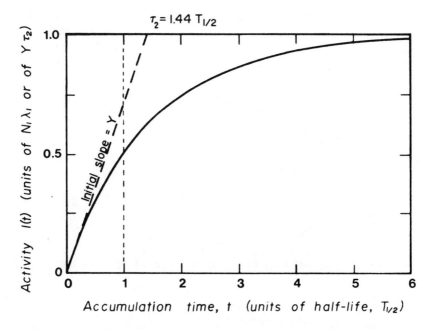

FIGURE 6.   Some relationships for radionuclide production from a nuclear reaction under conditions of steady bombardment. The exponential growth of the target activity towards its saturation value is evident.

$$I_2(t) = Y\tau_2(1 - e^{-\lambda_2 t}) \qquad\qquad (43)$$

The maximum activity which can be produced, $Y\tau_2$ is called the *saturation activity*. In the example above of 14 MeV protons bombarding ^{14}N, the saturation activity, $Y\tau_2$, would be 110 mCi/μA. This is also the *effective activity*, $N_1\lambda_1$, of the parent target while under bombardment. In one half-life, (20.4 min for ^{11}C), half of this saturation activity can be accumulated. Evidently it is inefficient to run the accelerator for very many half-lives because the rates of production and decay become nearly equal after a few half-lives ($e^{-\lambda t} = 0.03$ when $t = 4T_{1/2}$). Figure 6 illustrates some of the relationships involved in radionuclide production from nuclear reactions.

It should be noted that a particular target species, under bombardment, may undergo several competitive nuclear reactions and hence be the source of several daughter species. Because these different reactions are competitive, they may be treated independently. The yield of each reaction will depend on the conditions of bombardment.

4. General Equations of Radioactive Growth and Decay

Prior to this section we have dealt in considerable detail with the specific case of parent and daughter relationship in radioactive decay. Our approach could be extended to include a granddaughter product N_3, and so on, but rather than following that route, we will now present the general formalism governing any number of species.[6] Then our previous results for the parent-daughter relationships will be seen as a particular case of the general formalism, and any other particular cases can be readily extracted.

Given that at time $t = 0$ we have $N_1(0)$, $N_2(0)$, ...$N_n(0)$ atoms of radioactive species 1,2...n which are genetically related, so that each decays with decay constant $\lambda_1, \lambda_2 .... \lambda_n$ into the next species in the series, we wish to know the number of atoms $N_1(t)$, $N_2(t)$, ...$N_n(t)$ present at any subsequent time.

For each species the fundamental decay law $dN_i/dt = -\lambda_i N_i$ applies, but the number of atoms of each species beyond the first also increases because the decay of the preceding species in the series is continuously furnishing new atoms. Thus we can write

$$\frac{dN_1}{dt} = -\lambda_1 N_1$$

$$\frac{dN_2}{dt} = \lambda_1 N_1 - \lambda_2 N_2$$

$$\frac{dN_3}{dt} = \lambda_2 N_2 - \lambda_3 N_3$$

$$\vdots$$

$$\frac{dN_n}{dt} = \lambda_{n-1} N_{n-1} - \lambda_n N_n \qquad (44)$$

This is a system of differential equations which can be solved by setting

$$N_1 = A_{11}\, e^{-\lambda_1 t}$$

$$N_2 = A_{21}\, e^{-\lambda_1 t} + A_{22}\, e^{-\lambda_2 t}$$

$$N_3 = A_{31}\, e^{-\lambda_1 t} + A_{32}\, e^{-\lambda_2 t} + A_{33}\, e^{-\lambda_3 t}$$

$$\vdots$$

$$N_n = A_{n1}\, e^{-\lambda_1 t} + \ldots\ldots A_{nn} e^{-\lambda_n t} \qquad (45)$$

The constants A_{ij} are to be determined in such a way that the expressions of Equation 45 satisfy the differential Equation 44 and the $N_i(0)$ have the prescribed initial conditions. Substituting Equation 45 into Equation 44, we have

$$A_{ij} = A_{i-1,j}\, \frac{\lambda_{i-1}}{\lambda_i - \lambda_j} \qquad (46)$$

This recursion formula determines all the A_{ij} except those with equal indexes $i = j$. These are determined from the initial conditions

$$N_i(0) = A_{i1} + A_{i2} + \cdots A_{ii} \qquad (47)$$

Each of the expressions of Equation 45 involves a sum of exponentials containing the decay constants of all the species preceding the one considered.

The special case of initial conditions having only species 1 present at $t = 0$ with $N_1(0)$ atoms is of most importance. We have then, by direct application of Equations 45 through 47:

$$N_1(t) = N_1(0) e^{-\lambda_1 t}$$

$$N_2(t) = N_1(0)\, \frac{\lambda_1}{\lambda_2 - \lambda_1}\, (e^{-\lambda_1 t} - e^{-\lambda_2 t})$$

$$N_3(t) = N_1(0)\lambda_1\lambda_2 \left(\frac{e^{-\lambda_1 t}}{(\lambda_2 - \lambda_1)(\lambda_3 - \lambda_1)} + \frac{e^{-\lambda_2 t}}{(\lambda_3 - \lambda_2)(\lambda_1 - \lambda_2)} \right.$$

$$\left. + \frac{e^{-\lambda_3 t}}{(\lambda_1 - \lambda_3)(\lambda_2 - \lambda_3)} \right) \qquad (48)$$

We have previously derived the expressions here for $N_1(t)$ and $N_2(t)$; the expression for $N_3(t)$ represents the number of atoms at time t of a granddaughter product from parent stock $N_1(0)$ decaying through daughter $N_2(t)$. Expressions for further members in the series can be obtained through appropriate application of the Equations 45 through 47. The activity $I_i(t)$ of any product $N_i(t)$ is, of course, $N_i\lambda_i$.

The general Equation 45 provides the amount and activity at any time t of each radioactive product in a series, due to a nuclear irradiation or a radioactive accumulation. If the primary source of activity (or effective activity in the case of an irradiation) is removed, the amount and activity of each product at any subsequent time can also be calculated from Equation 45. The amount N_i, for example, remaining at any later time is made up of (1) a supply from $N_1 \ldots N_{i-1}$, each acting independently as a source and producing N_i in accordance with Equation 45; and (2) the residue of the original amount of N_i present, $N_i(0)$, which decays exponentially. Thus:

$$I_i = \left(\text{growth from } I_1(0)\right) + \left(\text{growth from } I_2(0)\right) + \ldots .$$

$$+ \left(\text{growth from } I_{i-1}(0)\right) + \left(\text{residue of } N_i(0)\right) \tag{49}$$

No matter how complicated the conditions of bombardment accumulation and decay this approach with the application of Equation 45 will yield the desired amount or activity.

5. Accumulation of Stable End Products

Radioactive decay series terminate sooner or later in a stable nuclide for which $\lambda = 0$. Usually the stable end products are not of much interest, as they emit no radiation and are present in such small quantities as to be usually not of chemical concern. Nonetheless, whenever an interest in stable end products does occur, the formulas in this chapter can be applied with the appropriate λ_i set equal to zero.

In the case of a parent N_1 and stable daughter N_2, then the application of Equation 22 with $\lambda_2 = 0$ yields

$$N_2(t) = N_1(0)\,(1 - e^{-\lambda_1 t}) = N_1(0) - N_1(t) = N_1(t)\,(e^{\lambda_1 t} - 1)$$

$$\tag{50}$$

These three equivalent expressions for $N_2(t)$ have been written out because they provide alternative (although of course equivalent) viewpoints.

The second form expresses the obvious relation $N_1(0) = N_1(t) + N_2(t)$; that is, the original $N_1(0)$ atoms, at any time t, either remain untransformed as $N_1(t)$ or have transformed into $N_2(t)$. The third form suggests that if the residual amount of $N_1(t)$ and of its decay product $N_2(t)$ are known, as well as the decay constant λ_1, then the time of accumulation t can be evaluated. This is the basis for geological age estimates from measurements of the accumulation of the stable end products [206]Pb or [208]Pb in uranium or thorium minerals.

The accumulation of a stable granddaughter product N_3 can be obtained from Equation 48 with $\lambda_3 = 0$:

$$N_3(t) = N_1(0) \left[1 - e^{-\lambda_1 t} - \frac{\lambda_1}{\lambda_2 - \lambda_1} (e^{-\lambda_1 t} - e^{-\lambda_2 t}) \right]$$

$$= N_1(0) - N_1(t) - N_2(t) \tag{51}$$

Thus, at time t, the original $N_1(0)$ atoms are divided between the residual $N_1(t)$, the daughter $N_2(t)$ and the stable product $N_3(t)$.

If the supply of active material is a nuclear reaction with effective target activity $N_1(0)\lambda_1$, where λ_1 is very small, it must be recognized that λ_1 should not be put equal to zero, but that the approximation $e^{-\lambda_1 t} = 1 - \lambda_1 t$ is valid. Thus for a stable product $N_3(t)$ from a reaction with target activity $N_1\lambda_1$ we obtain

$$N_3(t) = N_1\lambda_1 \left[t - \frac{1}{\lambda_2} (1 - e^{-\lambda_2 t}) \right]$$

which by substitution from Equation 37 (for $N_2(t) = (I_2(t)/\lambda_2)$) yields

$$N_3(t) = N_1(0)\lambda_1 t - N_2(t) \qquad (52)$$

The steady state production rate $N_1(0)\lambda_1$ throughout time t yields $N_2(t)$ radioactive atoms and $N_3(t)$ stable atoms.

6. Graphical Solutions

Most questions involving the amount or activity of a radioactive species will be quickly answered with application of the appropriate equations and a pocket calculator. Nevertheless, if a particular decay series is to be dealt with repeatedly, it may be advantageous to plot fractional activity values on a logarithmic scale, with an appropriate time axis as abscissa.

Suppose we are involved with the production of ^{131}I. The most common method involves the neutron irradiation of ^{130}Te in a reactor to yield ^{131}Te. This decays with a half-life of 30 hr to ^{131}I, which in turn decays with a half-life of 8.1 days to ^{131}Xe. The growth of ^{131}Te and ^{131}I in a ^{130}Te parent is plotted on a semilogarithmic scale in Figure 7, for the case of a constant irradiation of ^{130}Te with an effective rate $N_1\lambda_1 = 1$. In Figure 8 is plotted the activity (expressed as a fraction of I(0)) during the decay of ^{131}Te and growth of the daughter ^{131}I in an initially pure source of ^{131}Te.

If the irradiation conditions are such that the effective $N_1\lambda_1$ of ^{130}Te is 10 Ci, then by reference to Figure 7 we can determine the activity of ^{131}Te and of ^{131}I after any time of irradiation. Specifically, after 50 hr of irradiation there would be available 0.7 × 10 = 7 Ci of ^{131}Te and 0.07 × 10 = 0.7 Ci of ^{131}I. If this mixture were shipped to a radiopharmaceutical company which received the shipment exactly 4 days after the end of the irradiation, we can determine the activity of the ^{131}Te and ^{131}I upon arrival from Figure 8. But there is a problem; at the end of irradiation there was a mixture of ^{131}Te and ^{131}I in 10:1 activity ratio, whereas Figure 8 refers to the decay of an initially pure ^{131}Te sample. This dilemma can be overcome by noting from Figure 8 that after about 22 hr of decay, the pure ^{131}Te source becomes a mixture of ^{131}Te and ^{131}I, also with a 10:1 activity ratio. It must also be appreciated that the subsequent history of a mixture having any particular activity ratio is completely independent of how the mixture originated. Therefore, Figure 8 can be used to predict the activity of this particular 10:1 mixture at any later time if the time origin for the post-irradiation period is shifted to 22 hr. Then, 96 hr later, the ^{131}Te activity can be determined as 7 × 0.067/0.62 = 0.76 Ci, and the ^{131}I activity as 7 × 0.115/0.62 = 1.3 Ci. It is evident from the graph that this delivery time corresponds approximately to the time of maximum activity of ^{131}I from this particular irradiation sample.

7. Numerical Examples
a. ^{14}C Production and Decay

The production and decay of ^{14}C provides an interesting example of a "naturally" occurring radionuclide that is produced by a nuclear reaction. ^{14}C is produced by the reaction of cosmic rays with nitrogen and oxygen in the atmosphere. The production

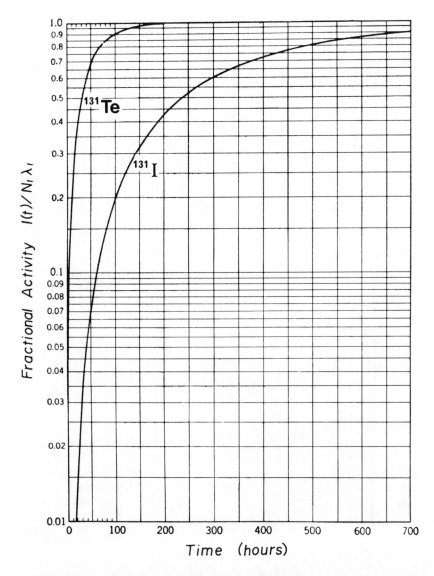

FIGURE 7. A semilogarithmic plot of the production of ^{131}Te and ^{131}I from the neutron irradiation of ^{130}Te. The use of such a graph to determine activity during irradiation is described in the text.

rate is equivalent to an effective activity $N_1\lambda_1$ of about 0.5 Ci/km^3 of atmosphere (sea level equivalent). The half-life of ^{14}C is 5570 years. The ^{14}C thus formed mixes with the stable carbon ^{12}C which is present as CO_2 in the atmosphere. If the world was created at time t = 0 with the same atmospheric conditions as now, and the same cosmic ray flux, then Equation 41 yields the growth of ^{14}C activity. After one half-life the activity has increased to 0.25 Ci/km^3, one half of its saturation value. Because the world is much older than this and the changes in atmospheric conditions are not very large over the time of a few half-lives, then we can assume that the saturation activity of ^{14}C (0.5 Ci/km^3) has been reached and the ^{14}C/^{12}C ratio in the atmosphere is constant. However, if some of this equilibrium mixture enters into a living organic structure (e.g., plant or animal) which subsequently dies and leaves the carbon locked in place, then the equilibrium is no longer maintained — the ^{14}C decays but is not replenished. If the mass of carbon in the atmosphere is 1.34×10^9 g/km^3, then the equilibrium saturation activity would correspond to $0.5/(1.3 \times 10^9)$ Ci or 0.4 nCi/g of carbon.

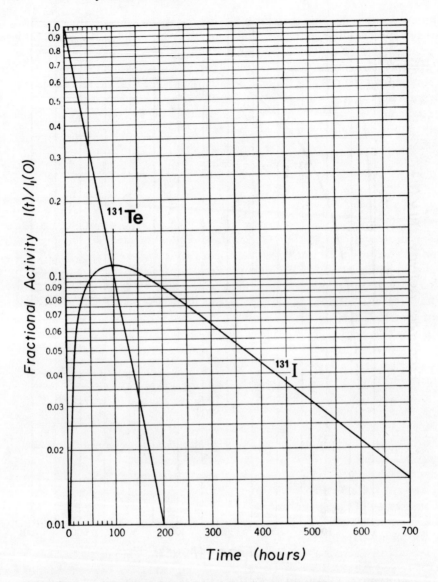

FIGURE 8.   A semilogarithmic plot of the decay of parent ¹³¹Te and daughter ¹³¹I.
This graph can be used to determine their activity at some time following their production, as described in the text.

If a bone fragment containing 10 g of carbon was unearthed 40,000 years after the
death of its owner, then its ¹⁴C activity could be estimated with reference to Equation
6 as I (40,000 years) = 0.4 e^{-5} nCi = 2.7 pCi. This is about the limit of sensitivity for
direct counting measurements even with excellent shielding against natural background. As is apparent, this is the basis of the ¹⁴C dating method, which considers
I(0) and λ as approximately known, and deduces t from a measurement of I(t).

b. ¹³¹I Decay

In the earlier example treating graphically the production and decay of ¹³¹Te and
¹³¹I, it was assumed that the product ¹³¹Xe is stable. This is not entirely true: 93% of
¹³¹I decays to the stable ¹³¹Xe ground state, 6.3% decays to a very short-lived (less than
10^{-9} sec) metastable state ¹³¹ᵐXe, but 0.7% decays to the metastable state ¹³¹ᵐ′Xe, which
has an 11.8 day half-life, longer than the 8.1 day half-life of ¹³¹I. For most purposes

this minor decay route is inconsequential, but it may become significant in long-term iodine retention studies. Assuming that pure ^{131}I is separated at t = 0, we can predict at what later time the activity of $^{131m'}$Xe will build up to equal the decaying ^{131}I activity. This example is an illustration of partial decay rates discussed in Section I.A.3, and in that section it was noted that the partial activity decayed with the same decay constant as the total activity, but with activity in the ratio of λ_1/λ to the total activity.

We proceed then to use Equation 32, the relation for the activity ratio of long-lived daughter to shorter-lived parent, remembering to reduce the daughter activity by the factor of 0.007. Substituting $\lambda_1 = 0.123$ day^{-1} and $\lambda_2 = 0.085$ day^{-1}, we obtain $I_2(t)/I_1(t) = 0.007 \times 2.2$ (e$^{0.038t}-1$), with t in units of days. The ratio I_2/I_1 will be unity after t = 110 days.

c. Impurities in ^{123}I

The iodine isotope ^{123}I is very attractive for medical application but there has been a serious problem of availability. One possible production method is a low energy (p,n) reaction on tellurium targets which have had their concentration of the ^{123}Te isotope enriched to about 20%. The problem with this approach is that due to the (p,n) reaction on ^{124}Te also in the target, there is an undesirable ^{124}I impurity in the final target. The half-lives for ^{123}I and ^{124}I are 13.2 hr and 100 hr, respectively. Suppose that at the end of an irradiation the fraction of ^{124}I impurity in ^{123}I is 1%. We can deduce the impurity fraction after any time, say 3 days. The initial activities are in the ratio of 1:100, but after 72 hr the ^{123}I is reduced by a factor of e$^{-3.8}$ = 0.022, whereas the ^{124}I is reduced by only a factor of e$^{-0.5}$ = 0.6. Thus after 3 days the ^{124}I/^{123}I ratio has increased to 27:100.

d. The ^{99}Mo − ^{99m}Tc Generator

The parent ^{99}Mo, half-life 66 hr, which decays to daughter ^{99m}Tc, half-life 6 hr, forms the most common radionuclide generator system in nuclear medicine laboratories. This example demonstrates some of the previously derived growth and decay relationships by obtaining: (1) the time to reach maximum ^{99m}Tc activity in the generator, and (2) the amount of ^{99}Tc present after periods of 1 day and 3 days from a previous elution of the generator.[8]

The metastable nuclide ^{99m}Tc decays to its ground state ^{99}Tc with a half-life of 6 hr. Thus the parent-daughter activity plot of Figure 4 represents the time scale of the ^{99}Mo − ^{99m}Tc system. There is one difference; only 86% of the ^{99}Mo decays to the metastable state ^{99m}Tc and the rest goes directly to the ground state ^{99}Tc. The equations of Section I.B.1 for the growth of a daughter product will need modification to fit this particular case. Inspection shows that Equations 17 through 23 require only the insertion of the factor 0.86 in front of N_1, and Equation 24 requires the factor 0.86 in front of I_1. In other words, the equations derived for the case of 100% decay to the daughter can be used to express the number and activity of the ^{99m}Tc daughter nuclide relative to its ^{99}Mo parent if the factor of 0.86 is inserted in front of the parent number and activity. Thus the daughter activity of Figure 4 needs to be multiplied by 0.86 to correctly represent the ^{99m}Tc activity. We also note that the decay product ^{99}Tc is not stable but has such a long half-life (2×10^5 years) that we shall neglect its decay.

The time t_m of maximum activity of a daughter product (starting with zero daughter at t = 0) is given in Equation 28. This equation requires no modification for the present example — the shape, and hence time of maximum of the daughter activity curve, has not changed. Substituting $\lambda_1 = 0.0105$ hr^{-1} and $\lambda_2 = 0.1155$ hr^{-1} in Equation 28, we obtain t_m = 23 hr.

The accumulation of ^{99}Tc in the generator column at time t following the previous (complete) eluding at time t = 0 can be readily estimated relative to the amount of

^{99m}Tc. This ratio is important because the ^{99}Tc can compete with the active ^{99m}Tc in subsequent chemical exchange.

The ^{99}Tc will be produced both through the direct decay of ^{99}Mo, with branching ratio 14%, and through the decay of the daughter ^{99m}Tc. With $N_1(0)$ atoms of ^{99}Mo at time t = 0, the growth of ^{99}Tc through the first route will be given by a modification of Equation 50:

$$N_3(t)_a = 0.14 \, N_1(0) \, (1 - e^{-\lambda_1 t})$$

and the accumulation through the second route will be given by a modification of Equation 51:

$$N_3(t)_b = 0.86 \, N_1(0) \left[1 - e^{-\lambda_1 t} - \frac{\lambda_1}{\lambda_2 - \lambda_1} \, (e^{-\lambda_1 t} - e^{-\lambda_2 t}) \right]$$

The sum of these two branches yields an accumulation of ^{99}Tc equal to:

$$N_3(t) = N_1(0) \left[1 - e^{-\lambda_1 t} \right] - 0.86 \, N_1(0) \left[\frac{\lambda_1}{\lambda_2 - \lambda_1} \, (e^{-\lambda_1 t} - e^{-\lambda_2 t}) \right]$$

For the accumulation of the daughter ^{99m}Tc, we have a modification of Equation 22:

$$N_2(t) = 0.86 \, N_1(0) \, \frac{\lambda_1}{\lambda_2 - \lambda_1} \, (e^{-\lambda_1 t} - e^{-\lambda_2 t})$$

Thus the ratio of "stable" ^{99}Tc atoms to active ^{99m}Tc atoms at any time t is given by the ratio $N_3(t)/N_2(t)$ substituted from the above equations. At time t = 24 hr the ^{99}Tc/^{99m}Tc ratio, $N_3(24)/N_2(24)$, equals 2.6. At time t = 72 hr, the ratio has increased to 12.2. Obviously, any disadvantageous chemical effects from the inactive ^{99}Tc component can be reduced by avoiding long periods between eluding the generator.

II. STATISTICAL FLUCTUATIONS IN RADIOACTIVE DECAY

The first half of this chapter considered radioactive decay from the point of view of a continuous change in the number of atoms. This approach facilitated the mathematical formalism for the presentation of the exponential growth and decay laws. Very often the results from the continuum theory are entirely adequate for the need at hand; as mentioned previously, the continuum theory is a good approximation for processes involving a large number of atoms and it yields exactly the average behavior of a large number of initially identical ensembles of radioactive atoms.

However useful the continuum theory, the fundamental nature of radioactive decay is random. There will be occasions when it is important, or useful, to know not only the average value m of some parameter such as the number of decaying atoms, but also to estimate the distribution of actual values x about the average. The remainder of this chapter will be devoted to a discussion of the statistical fluctuations in radioactive decay. More intensive and extensive treatments can be found in monographs.[9-11] The book by Evans[12] contains a chapter on the application of statistics to counting instruments and an appendix on inefficient statistics. We do not discuss the use of inefficient statistics (simplified formulas which extract less than the maximum amount of available information from the data) because the ease of electronic calculations renders them less attractive.

A. Frequency Distributions

In any series of measurements, the frequency of occurrence of particular values is expected to follow some *frequency distribution*. There are several distributions which are relevant to the discussion and measurement of radioactive events. Before considering them individually we will discuss some general aspects and definitions regarding frequency distributions.

A *frequency distribution* will be considered as any graph of the measured or predicted frequency of obtaining particular values of some parameter. It compactly represents the measured or predicted variations in successive observations. Frequency distributions may be symmetrical or nonsymmmetrical. They may be discrete (a histogram) or continuous (a smooth curve). They may have a finite range or an infinite ($-\infty$ to $+\infty$) or semi-infinite (0 to $+\infty$) range.

A *probability distribution* can be regarded as a special case of a normalized frequency distribution such that its sum (if a discrete distribution) or its integral (if continuous) is equal to unity.

In order to describe a particular frequency distribution, we may use (1) a *location index* of the center of the distribution, and (2) a measure of the spread or dispersion of the distribution, referred to as a *precision index*.

1. Location Indexes

The three commonly used location indexes which describe the center of a frequency distribution are the *median*, the *mode*, and the *mean*. When a distribution has a single maximum and is symmetrical, these three location indexes are all identical in value. But because many frequency distributions are asymmetrical, we shall define the three location indexes.

The *median* is the middle measurement of an odd number of measurements (ordered as to magnitude); or the interpolated middle value. The median is used on occasion to bypass part of the effect of a strongly skewed distribution, but it is not very important in experimental science.

The *mode* is the most probable value, or the abscissa value corresponding to the peak of the frequency distribution. It is of interest in asymmetrical distributions, particularly with mathematical model distributions. It is of much less interest for actual measurements, because it is difficult to determine accurately from experimental data.

The *mean value* (or arithmetic average) is by far the most important location index. We shall define an experimental mean m, such that for n equally weighted trial measurements, $x_1, x_2, \ldots x_n$

$$m = \frac{\sum_{i=1}^{n} x_i}{n} \qquad (53)$$

If values x_i are observed more than once, say f_i times, then the different values of x_i can be weighted by the corresponding frequency of occurrence f_i;

$$m = \frac{\sum_{i=1}^{n'} f_i x_i}{n} \qquad (54)$$

where the sum is now over the n' (n' < n) different x values in the n trials. The experimental probability for observing x_i can be denoted $p_i = f_i/n$, so that Equation 54 becomes

$$m = \sum_{i=1}^{n'} x_i p_i \qquad (55)$$

It is also useful to define a hypothetical mean for an infinite number of measurements, or for a mathematical model. In parallel with Equation 55, and depending on whether we are dealing with a discrete or continuous parent distribution, we have:

$$\mu = \sum_{i=0}^{\infty} x_i p_i \quad \text{or} \quad \mu = \int_0^{\infty} x p(x)\, dx \qquad (56)$$

where $p(x)$ is introduced as the continuous probability function over the range of sample space from 0 to ∞.

The mean μ is sometimes called the *parent* mean and m the *sample* mean (or estimator of the parent). The value of m will fluctuate in different sets of measurements, but will approach the value μ as n is made larger and larger. A frequent objective in the statistical interpretation of measurements is the best determination of μ, not of m, but generally the best available value of μ from a given number of trials n is the value of m obtained from Equation 53. In speaking of the mean of a mathematical model distribution, we refer to μ. Once convinced that a model fits the experimental measurements, we approximate μ by m and take advantage of further predictions from the model.

As defined above, the mean is the abscissa value of the center of area (the centroid) of a frequency distribution. It is also referred to as the expectation value. Alternative means (e.g., the geometric) may be defined, but will not be used here.

2. Dispersion Indexes

Dispersion indexes are measures of the spread or dispersion of a frequency distribution about the central value. The dispersion is thus a measure of precision. Two possible dispersion indexes which we shall mention only in passing are the range and quantiles. They do not provide sufficiently quantitative information to be very useful to us. The most useful dispersion indexes are the *moments* of a frequency distribution about the mean. Before defining the moments, we should discuss the simpler dispersion indexes; the *deviation, mean deviation, standard deviation,* and *variance.*

The *deviation* (or statistical fluctuation) from the mean, Δx_i, for the ith measurement of a set of n trials is simply

$$\Delta x_i = x_i - m \qquad (57)$$

It follows that $\sum_{i=1}^{n} \Delta x_i = 0$.   It is convenient to define the deviation with reference to the mean because then the sum of the squares of all the deviations is a minimum. This is readily verified. If S denotes the sum of the squares of the deviations $\Delta x'$ defined with reference to some unspecified reference value m', then

$$S = \sum_{i=1}^{n} \Delta x_i'^2 = \sum_{i=1}^{n} (x_i - m')^2 = \sum_{i=1}^{n} x_i^2 - 2m' \sum_{i=1}^{n} x_i + n m'^2$$

$$(58)$$

To find the particular value of m' (call it m'') that makes S a minimum, we differentiate S with respect to m' and place the derivative equal to zero.

$$\left(\frac{dS}{dm'}\right)_{m'=m''_n} = -2\sum_{i=1}^{n} x_i + 2nm'' = 0$$

$$m'' = \frac{\sum_{i=1}^{n} x_i}{n} \tag{59}$$

This is, of course, our original definition of the mean (Equation 53) so that we have verified that m is the reference value for which the sum of the squares of the experimental deviations is a minimum. When we deal with a mathematical model distribution, the reference value for S to be a minimum is the parent mean μ.

The *mean* (or average) *deviation* is defined without regard to the algebraic signs of the individual deviations. Thus the mean deviation, $\overline{\Delta x}$, is given as

$$\overline{\Delta x} = \frac{\sum_{i=1}^{n} |\Delta x_i|}{n} \tag{60}$$

The mean deviation is not as efficient an index as the standard deviation, so that its use is not to be encouraged.

The *standard deviation* is another measure of the dispersion of a frequency distribution. For a limited number n of equally weighted measurements, the experimental standard deviation s is defined as

$$s = \left(\frac{\sum_{i=1}^{n} (x_i - m)^2}{n}\right)^{1/2} = \left(\frac{\sum_{i=1}^{n} \Delta x_i^2}{n}\right)^{1/2} \tag{61}$$

For the case of x_i observed f_i times, this definition can also be written:

$$s = \left(\frac{\sum_{i=1}^{n'} \Delta x_i^2 f_i}{n}\right)^{1/2} = \left(\sum_{i=1}^{n'} \Delta x_i^2 p_i\right)^{1/2} \tag{62}$$

The standard deviation, s, is also called the *root-mean-square* (or rms) deviation. Because the deviations in Equation 61 are individually squared before the summation is made, more weight is assigned to the large deviations. Hence, as a measure of dispersion, s is more sensitive to large deviations than is the mean deviation $\overline{\Delta x}$. Because of the squaring operation involved in its definition, the standard deviation does not allow distinction between the algebraic sign of individual deviations and gives no indication of the degree of symmetry of the distribution. The standard deviation provides, as does any dispersion index, a numerical guess as to the likely range of values into which the next measurement may fall. With this interpretation, s is sometimes referred to as the standard deviation of a single measurement, rather than of the distribution.

The parent standard deviation, which is based on the reference mean μ rather than m for computing each deviation, will be assigned the name σ. Its square, σ^2, is called the *variance*. In the two cases of a discrete and a continuous parent distribution, we have

$$\sigma^2 = \sum_{i=0}^{\infty} (x_i - \mu)^2 p_i \quad \text{and} \quad \sigma^2 = \int_0^{\infty} (x - \mu)^2 p(x) dx \qquad (63)$$

where we have again used $p(x)$ as the (normalized) continuous probability function over the range from 0 to ∞. The variance σ^2 is statistically the most important parameter in describing the dispersion of any parent distribution, including any mathematical model distribution. Strictly speaking, the relationship of Equation 63 indicates that μ must be known in order to calculate σ^2, but the value of μ, and hence of σ^2, can never be exactly determined from any finite set of measurements. We must use the best estimate for μ that can be obtained. As mentioned previously, the best value of μ is generally taken as the experimental value m; the best approximation to σ^2 that can be deduced from a set of n measurements is generally taken as

$$\sigma^2 \simeq \frac{n}{n-1} s^2 \qquad (64)$$

Why is the best estimate of σ^2 not simply s^2? The best estimators cannot be defined without reference to additional conditions or criteria, such as the maximum likelihood criteria. A discussion of these estimates is given by Bacon.[13] The choice specified above will not be justified here, but it can be pointed out in passing that the $n/(n-1)$ factor in Equation 64 is related to the degrees of freedom of n measurements — there are only $n-1$ independent deviations.

Although the concept of the difference between s and σ (and between m and μ) is very important, in practice these differences are usually insignificant. Neither s nor σ is numerically very significant unless n is reasonably large, and then the differences between m and μ, and between n and $n-1$ are small.

Some of the terms we have been defining are related to the *moments* of a frequency distribution, so that it is convenient to introduce a general definition. The *kth moment* about the *origin*, θ_k^o, of a frequency distribution of n measurements in which x_i occurs with frequency f_i is

$$\Theta_k^o = \frac{\sum_{i=1}^{n'} x_i^k f_i}{n} = \sum_{i=1}^{n'} x_i^k p_i \qquad (65)$$

where $n' \leqslant n$ is the number of different x_i and p_i is the probability of occurrence of x_i. It is apparent from a comparison with Equation 55 that the mean is simply the first moment about the origin,

$$\Theta_1^o = m \qquad (66)$$

The *kth moment* about the *mean* m, θ_k^m, is

$$\Theta_k^m = \frac{\sum_{i=1}^{n'} (x_i - m)^k f_i}{n} = \sum_{i=1}^{n'} (x_i - m)^k p_i \qquad (67)$$

By comparison with Equation 62, it is seen that the square of the standard deviation corresponds to the second moment about the mean of a frequency distribution:

$$\Theta_2^m = s^2 \tag{68}$$

It should be noted that the first moment about the mean, θ_1^m, does not correspond to the mean deviation $\overline{\Delta x}$. In fact, because of the definition of the mean, θ_1^m has value zero, and this is why $\overline{\Delta x}$ is defined without regard for algebraic sign.

A relationship exists between the second moments about the origin and about the mean that is sometimes useful for calculations. Thus

$$\Theta_2^m = \sum_{i=1}^{n'} (x_i - m)^2 p_i = \sum_{i=1}^{n'} x_i^2 p_i - 2m \sum_{i=1}^{n'} x_i p_i + m^2 \sum_{i=1}^{n'} p_i$$

$$= \Theta_2^o - 2m\Theta_1^o + m^2$$

and by substituting from Equation 66,

$$\Theta_2^m = \Theta_2^o - m^2 \tag{69}$$

Higher moments about the mean are also used as dispersion indexes. As evident from the definition in Equation 67, the higher the moment about the mean the greater is the relative weighting of the large deviations. A dispersion index called the *coefficient of skewness* gives a measure of the asymmetry of a frequency distribution, and it is defined in terms of the third moment about the mean.

$$\text{skewness (experimental)} = \frac{\sum_{i=1}^{n} (x_i - m)^3 p_i}{s^3} \tag{70}$$

$$\text{and skewness (parent)} = \int_0^\infty \frac{(x - \mu)^3 p(x) dx}{\sigma^3} \tag{71}$$

It is possible to use still higher moments as dispersion indexes, but their usefulness is very restricted for experimental distributions. Unless the number of measurements n is very large, the experimental higher moments are not very reliable because they are so sensitive to fluctuations in the tail regions of the distribution.

The *standard deviation in the mean*, s_m, is the final dispersion index which we will consider. It is also called the *standard error*. The mean value m and the standard deviation s were defined previously (Equations 53 and 61) with reference to a set of n measurements or trials. If a *second* set of n measurements were recorded, then a somewhat different value of the mean would be obtained. Indeed, if the set of n measurements were repeated many times, a frequency distribution of the mean values could be obtained. It is frequently useful to be able to determine the standard deviation in the mean. Of course, if a set of n measurements are repeated many times, say N, then the standard deviation of the mean, denoted s_m, can be determined by a straightforward application of Equation 53 to the N mean values:

$$s_m = \left(\frac{\sum_{j=1}^{N} (m_j - \overline{m})^2}{N} \right)^{1/2} \tag{72}$$

where $\overline{m}$ is the grand mean of the N mean values. But very often it is not possible or appropriate to perform the N × n measurements required to apply this definition. Fortunately, statistical theory provides an estimate of the standard deviations of the mean, s_m, from the n measurements of a single set. The result only will be quoted now and the derivation postponed until later (Equation 141). For a set of n measurements, having a mean value s, the standard deviation in the mean is

$$s_m = \frac{s}{\sqrt{n}} \tag{73}$$

If we refer to a parent or model distribution, then the formulas corresponding to Equations 72 and 73 are

$$\sigma_m = \lim_{N\to\infty} \left(\frac{\sum_{j=1}^{N} (m_j - \mu)^2}{N} \right)^{1/2} \text{ and } \sigma_m = \frac{\sigma}{\sqrt{n}} \tag{74}$$

Very often we may wish to determine the best value of the parent standard deviation in the mean, σ_m, when μ is not known and we have only an experimentally determined approximation of the mean, m. Then, in parallel with Equation 64, the best approximation to the parent standard deviation in the mean is given by

$$\sigma_m \simeq \frac{s}{\sqrt{n-1}} = \left(\frac{\sum_{i=1}^{n} (x_i - m)^2}{n(n-1)} \right)^{1/2} \tag{75}$$

with the decrease in the degrees of freedom occurring as s replaces σ.

The fractional standard deviation in the mean is a simple and useful entity. It is

$$\text{fractional } s_m = \frac{s_m}{m} = \frac{s}{m\sqrt{n}}$$

$$\text{fractional } \sigma_m = \frac{\sigma_m}{\mu} = \frac{\sigma}{\mu\sqrt{n}} \tag{76}$$

B. Mathematical Models of Frequency Distributions

We now turn to a discussion of the mathematical models of the frequency distributions which are most relevant to a discussion of radioactive decay. In this section many of the concepts introduced in a general fashion in the preceding section will be made more specific by applying them to particular frequency distributions.

1. The Binomial Distribution

The *binomial distribution* is the fundamental frequency distribution governing random events. It was the first probability distribution to be enunciated theoretically — by Bernoulli in the early 18th century. Suppose we repeat many times a certain experiment that must have one of two results, E or F, which are mutually exclusive. The probability of E occurring in each trial is p; the probability of F occurring is $1 - p = q$. What is the probability, in n trials, that E occurs k times and hence that F occurs n − k times? (The Bernoulli problem.)

Because the probability of E occurring in any one of the n trials is p, and the probability of F occurring in any specified trial is q, then the probability of E occurring in k particular trials and F occurring in n − k particular trials is

$$p^k q^{n-k} \tag{77}$$

But if we abandon the condition that E and F occur in particular trials and retain only the requirement that they must occur, then we must sum Equation 72 for all possible choices of the trials. These are equal in number to the combinations of n objects taken k at a time, denoted by $\binom{n}{k}$:

$$\binom{n}{k} = \frac{n(n-1)\cdots(n-k+1)}{n!} = \frac{n!}{k!\,(n-k)!} \tag{78}$$

with the factorial n, n!, having the usual meaning of n(n − 1) (n − 2)...(1). Thus the answer to the Bernoulli problem is

$$P_{n,p}(k) = \binom{n}{k} p^k q^{n-k} \tag{79}$$

Turning back another two centuries, we write Newton's binomial expansion for p + q:

$$(p+q)^n = \binom{n}{0}p^n + \binom{n}{1}p^{n-1}q + \binom{n}{2}p^{n-2}q^2 +$$
$$\cdots + \binom{n}{k}p^k q^{n-k} + \cdots + \binom{n}{n}q^n \tag{80}$$

It is seen that $P_{n,p}(k)$ is the term containing $p^k q^{n-k}$ in the binomial expansion. Hence

$$\sum_{k=0}^{n} P_{n,p}(k) = (p+q)^n = 1 \tag{81}$$

showing that the probabilities $P_{n,p}(k)$ are normalized correctly.

A graph of $P_{n,p}(k)$ vs. k is called the *binomial distribution* of probabilities. Examples are shown in Figure 9 for three particular pairs of values of the independent parameters p and n. The distribution is asymmetric except for the special case of p = ½, but the asymmetry reduces as the product np increases.

Now that we have obtained a model probability distribution, we can apply the formulas of the previous section for location and dispersion indexes. In particular, the *mean* and *variance* of the binomial model distribution can be written as the respective moments. Using Equation 79 for p_i in Equation 65, we have

$$\mu = \Theta_1^0 = \sum_{k=0}^{n} k\,\frac{n!}{k!(n-k)!}\,p^k q^{n-k} \tag{82}$$

Because of the factor k in the numerator, the sum is not altered by changing the lower limit from k = 0 to k = 1. If n and p are factored out, the expression becomes

$$\mu = np \sum_{k=1}^{n} \frac{(n-1)!}{(k-1)!\,(n-k)!}\,p^{k-1} q^{n-k} \tag{83}$$

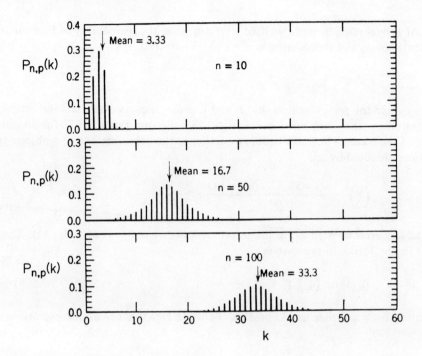

FIGURE 9.   Three examples of the binomial frequency distribution, $p_{n,p}(k)$. The parameter p is fixed at ⅓ for all three cases, with n taking the value 10, 50, and 100. Values of the mean, μ = np, are indicated.

Substituting $k' = k - 1$, and $n' = n - 1$, then

$$\mu = np \sum_{k'=0}^{n'} \frac{n'!}{(n' - k')!} \, p^{k'} q^{n'-k'} \tag{84}$$

But the summation is $\Sigma P_{n',p}(k')$, which, by Equation 81, is unity. Hence the parent mean of the binomial distribution is

$$\mu = np \tag{85}$$

The variance of the binomial distribution can be written, with the help of Equation 69, as

$$\sigma^2 = \Theta_2^\mu = \Theta_2^0 - \mu^2 = \sum_{k=0}^{n} k^2 \, \frac{n!}{k!(n-k)!} \, p^k q^{n-k} - \mu^2 \tag{86}$$

Replacing k^2 by $k(k - 1) + k$ and using Equation 84 yields

$$\sigma^2 = \sum_{k=0}^{n} k(k - 1) \, \frac{n!}{k!(n-k)!} \, p^k q^{n-k} + \mu - \mu^2$$

Because of the factor $k(k - 1)$, the summation is the same starting from $k = 2$. Factoring out $n(n - 1)p^2$ yields

$$\sigma^2 = n(n-1)p^2 \sum_{k=2}^{n} \frac{(n-2)!}{(k-2)!\,(n-k)!} \, p^{k-2} q^{n-k} + \mu - \mu^2$$

Substituting $k' = k - 2$ and $n' = n - 2$, then

$$\sigma^2 = n(n-1)p^2 \sum_{k'=0}^{n'} \frac{n'!}{(n'-k')!} p^{k'} q^{n'-k'} + \mu - \mu^2$$

and, again, the summation $\Sigma P_{n',p}(k') = 1$, so that

$$\sigma^2 = n(n-1)p^2 + \mu - \mu^2$$

Substituting np for μ from Equation 80 yields

$$\sigma^2 = np(1 - p) = npq \tag{87}$$

The fractional standard deviation, σ/μ can be easily obtained:

$$\frac{\sigma}{\mu} = \frac{[np(1-p)]^{1/2}}{np} = \left(\frac{1-p}{np}\right)^{1/2} = \left(\frac{1}{\mu} - \frac{1}{n}\right)^{1/2} \tag{88}$$

But what does all of this have to do with radioactive decay? The binomial is a very important distribution because many phenomena can be analyzed in terms of basic Bernoulli trials. It contains the two independent parameters n and p and rigorously applies to all phenomena in which the number of trials n and number k of occurrences of E are integers. It therefore describes the fluctuations in radioactive decay, with one proviso. The probability that a particular atom with decay constant λ will decay in observation time Δt is $\lambda \Delta t$. This decay can represent the event E occurring with probability p in the binomial distribution, provided that the mean life of the radioactive substance is sufficiently long that its activity does not change appreciably during the observations (so that p is constant). In such an approximation, $P_{n,\lambda\Delta t}(k)$ is the probability of k decays in n intervals, or in time $n\Delta t$.

Rather than pursue a discussion of the fluctuations in radioactive decay as predicted by the binomial distribution, it is advantageous to consider another probability distribution which is an approximation to the binomial distribution and which permits simpler calculations.

2. The Poisson Distribution

The *Poisson distribution* may be considered an approximation to the binomial distribution previously discussed. The conditions for the Poisson approximation are the following:

1. The number of trials n very large, infinite in the limit
2. The probability of occurrence p very small, zero in the limit
3. The mean value $\mu = np$ moderate in magnitude, in that $np \ll \sqrt{n}$

These conditions imply that, on the average, many trials are necessary before the event E occurs. For this reason the Poisson distribution is sometimes called the formula for the probability of rare events. The Poisson distribution is of interest here because its conditions for validity are satisfied by radioactive decay; it provides a convenient formalism for discussing the fluctuations of radioactive decay.

To derive the Poisson formula, we start with the binomial probability distribution as expressed in Equation 79 and recall that $p + q = 1$ and $\mu = np$. Thus Equation 79 can be written as

$$P_{n,p}(k) = \frac{1\left(1 - \frac{1}{n}\right)\left(1 - \frac{2}{n}\right)\cdots\left(1 - \frac{k-1}{n}\right)}{\left(1 - \frac{\mu}{n}\right)^k} \frac{\mu^k}{k!} \left(1 - \frac{\mu}{n}\right)^n$$

(89)

Under the Poisson conditions of n very large and $\mu \ll \sqrt{n}$, the first factor of Equation 89 is essentially unity. In this limit of large n, the last factor of Equation 89 can be approximated by an exponential, for

$$e^{-\mu} = 1 - \frac{\mu}{1!} + \frac{\mu^2}{2!} - \frac{\mu^3}{3!} + \cdots = \lim_{n \to \infty} \left(1 - \frac{\mu}{n}\right)^n$$

(90)

With these approximations, the binomial probability $P_{n,p}(k)$ becomes the Poisson probability,

$$P_\mu(k) = \frac{\mu^k e^{-\mu}}{k!}$$

(91)

It is noteworthy that, in contrast with the binomial distribution, this Poisson frequency distribution has only one parameter μ. It is a discrete distribution because k assumes only integer values. That the Poisson distribution of Equation 91 is properly normalized can be verified by a summation over all k:

$$\sum_{k=0}^{\infty} P_\mu(k) = e^{-\mu} \sum_{k=0}^{\infty} \frac{\mu^k}{k!} = e^{-\mu} e^{\mu} = 1$$

(92)

The Poisson distribution has been derived above as a special case of the binomial distribution. It may be helpful to our appreciation of its underlying principles, especially its relationship to the case of nuclear decay, to derive the Poisson frequency distribution from first principles.

The general conditions of validity for the Poisson distribution can be specialized to the specific case of radioactive disintegration of a group of atoms with the following necessary and sufficient conditions:

1. The probability of disintegration in a particular time interval is the same for all atoms in the group.
2. All atoms in the group are independent.
3. The probability of disintegration is the same for all time intervals of equal size (mean life long compared with the total period of observation).
4. The number of radioactive atoms and the number of equal time intervals are large.

If the average rate of disintegrations from the random decay of the group of atoms is a, then in the short time interval dt, such that a dt $\ll$ 1, the quantity adt is the probability P(1) of observing one disintegration in the time dt. The time interval dt can be made sufficiently small that the probability of observing two or more disintegrations in the time dt becomes vanishingly small in comparison with the probability of observing one particle. The probability of observing no particle in dt will then be $P_{dt}(0) = 1 - P_{dt}(1)$. The probability of observing k particles in the time (t + dt) may be written as the combined probabilities of (k − 1) particles in t and one in dt, and of k particles in t and none in dt; thus

$$P_{t+dt}(k) = P_{dt}(1)\ P_t\ (k-1) + P_{dt}(0)\ P_t(k)$$

$$= adt\ P_t(k-1) + (1-adt)\ P_t(k) \tag{93}$$

Rewriting in differential form,

$$\frac{d\ P_t(k)}{dt} = \frac{P_{t+dt}(k) - P_t(k)}{dt} = a[P_t(k-1) - P_t(k)] \tag{94}$$

The solution of this differential equation has been given as:[14]

$$P_t(k) = \frac{(at)^k}{k!}\ e^{-at} \tag{95}$$

which can be verified by differentiation. If the equal time intervals are chosen of length t, then the mean number of particles per interval is at $= \mu$, so that Equation 95 becomes

$$P_\mu(k) = \frac{\mu^k}{k!}\ e^{-\mu} \tag{96}$$

in agreement with Equation 91. $P_\mu(k)$ is the probability of observing k events (or in this specific case, disintegrations) when the average for a large number of trials (the parent mean) is μ.

A special case of Equation 91 is of particular interest. The probability of observing no event (k = 0, and noting that 0! equals 1, for n! = n(n − 1)!) is simply

$$P(0) = e^{-\mu} \tag{97}$$

This is the statistical basis for the law of radioactive decay, previously derived in Section I.A. The exponential $e^{-\mu t}$ is the probability that an atom will survive without decay for a time t when on the average it should have decayed $\mu = \lambda dt$ times.

The location and dispersion indexes defined earlier in Section II.A can be easily obtained for the Poisson frequency distribution.

The parent arithmetic mean, μ, was shown equal to np for the binomial distribution (Equation 85). Because of the definition of the Poisson distribution as an approximation to the binomial distribution, we can assume (as in the substitution for Equation 44) that in the Poisson case $\mu = np$ is also the arithmetic mean. But we can also derive this from the basic definition of Equation 56. Thus the mean of the Poisson frequency distribution (Equation 91) can be expressed as:

$$\lim_{n\to\infty} \sum_{k=0}^{n} kP_\mu(k) = e^{-\mu} \lim_{n\to\infty} \sum_{k=0}^{n} \frac{k\mu^k}{k!} = e^{-\mu}\mu \lim_{n\to\infty} \sum_{k=1}^{n} \frac{\mu^{k-1}}{(k-1)!} = e^{-\mu}e^{\mu}\mu = \mu \tag{98}$$

as expected.

We have seen that the standard deviation $\sigma = \sqrt{npq}$ for the case of the binomial distribution (Equation 87). Hence for the Poisson distribution, $\sigma = \sqrt{np}$ since $q \simeq 1$. Substituting the expression for the mean, $\mu = np$, yields

$$\sigma = \sqrt{\mu} \tag{99}$$

This expression can also be derived from the basic definition (Equation 63) of the variance σ^2. By expansion and substitution we obtain:

$$\sigma^2 = \lim_{n \to \infty} \sum_{k=0}^{n} (k - \mu)^2 P_\mu(k) = \lim_{n \to \infty} \sum_{k=0}^{n} k^2 P_\mu(k) - \mu^2 \tag{100}$$

Further manipulation, and a change of lower limit in the summation from 0 to 2 because of a factor of $k(k - 1)$ in the numerator, yields

$$\sigma^2 = \mu^2 \lim_{n \to \infty} \sum_{k=2}^{n} \frac{\mu^{k-2} e^{-\mu}}{(k - 2)!} + \mu - \mu^2 \tag{101}$$

The sum here is unity because of the normalization property, so that

$$\sigma^2 = \mu^2 + \mu - \mu^2 = \mu$$

and

$$\sigma = \sqrt{\mu} \tag{102}$$

The fractional standard deviation, where the standard deviation is expressed as a fraction of the mean, is very commonly used. It is

$$\text{fractional } \sigma = \frac{\sigma}{\mu} = \frac{1}{\sqrt{\mu}} \tag{103}$$

Some numerical examples in Section II.B.6 will show the utility of these expressions.

The Poisson distribution is asymmetric. As will be seen, its asymmetry or skewness decreases as the expectation value μ increases. By definition, the third moment about the mean is

$$\sum_{k=0}^{n} (k - \mu)^3 P_\mu(k) = \sum (k^3 - 3k^2\mu + 3k\mu^2 - \mu^3) P_\mu(k)$$

$$= \sum k^3 P_\mu(k) - 3\mu \sum k^2 P_\mu(k) + 3\mu^3 \sum k P_\mu(k) - \mu^3 \tag{104}$$

The quantity $k(k - 1)(k - 2)$ can be separated out of the first term and the summation begun at $k = 3$. Using also the normalization property, the first term becomes

$$\mu^3 - 3 \sum k^2 P_\mu(k) - 2 \sum k P_\mu(k)$$

Substituting $\sum k^2 P_\mu(k) = \sigma^2 + \mu^2$ (Equation 100), $\sigma^2 = \mu$ (Equation 102) and $\sum k P_\mu(k) = \mu$ (Equation 98) into Equation 104, we get for the third moment about the mean

$$\sum_{k=0}^{n} (k - \mu)^3 P_\mu(k) = \mu \tag{105}$$

The skewness by definition is the third moment about the mean divided by σ^3, so that

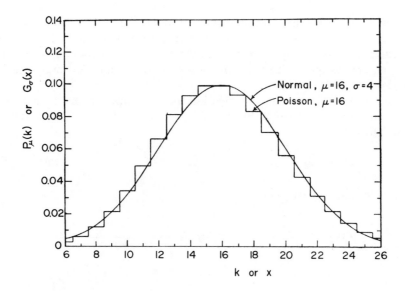

FIGURE 10. The Poisson and normal frequency distribution, both with μ = 16. The Poisson distribution is noticeably asymmetric. For the normal distribution, the parameter σ is assigned the value 4 in order to match the dispersion of the Poisson distribution.

$$\text{skewness} = \sum_{k=0}^{n} \frac{(k - \mu)^3 P_\mu(k)}{\sigma^3} = \frac{\mu}{\mu^{3/2}} = \frac{1}{\sigma} \qquad (106)$$

Because of the skewness of the Poisson distribution the most probable k value, k_0, or mode of the distribution does not correspond exactly with the mean μ. From the expression for the Poisson distribution $P_\mu(k)$ in Equation 91 it readily follows that

$$\frac{P_\mu(k + 1)}{P_\mu(k)} = \frac{\mu}{k + 1} \qquad (107)$$

From the definition of the most probable value k_0 we can write the inequalities

$$P_\mu(k_0 + 1) \leqslant P_\mu(k_0) \quad \text{and} \quad P_\mu(k_0 - 1) \leqslant P_\mu(k_0) \qquad (108)$$

Substituting Equation 107 into these inequalities yields

$$\mu - 1 \leqslant k_0 \leqslant \mu \qquad (109)$$

showing that the most probable value of k_0 is not more than one unit of k below the mean value μ.

The Poisson distribution $P_{16}(k)$ for $\mu = 16$ is plotted for illustration in Figure 10.

3. The Normal Distribution

The *normal probability distribution* (also called the Gaussian distribution, although Laplace and De Moivre both derived it many years before Gauss) is an analytical approximation to the binomial distribution, valid for large n and moderate p (np $\gg$ 1). It is very important in statistics and in the analysis of errors, but in the discussion of radioactive decay it is not as important as the Poisson distribution.

When the variable n of the binomial distribution becomes very large, the unit interval between adjacent k values becomes small relative to the mean value μ = np. Thus the distribution tends towards a continuous one. Rather than derive the normal probability distribution as an approximation to the binomial distribution, we shall derive it here as an approximation to the Poisson distribution, valid for large n (infinite in the limit) and $\mu \gg 1$.

We define a variable z in terms of deviation from the mean value u or most probable value k_0. For large μ, the small difference between u and k_0 (Equation 108) will be neglected. Thus

$$z = k - k_0 \approx k - \mu \tag{110}$$

Substituting in Equation 91 yields

$$P_\mu(k_0 + z) = \frac{\mu^{k_0 + z} e^{-\mu}}{(k_0 + z)!} = P_\mu(k_0) \; \frac{\mu^z}{(k_0 + 1)(k_0 + 2) \ldots (k_0 + z)}$$

or

$$\ln P_\mu(k_0 + z) = \ln P_\mu(k_0) + z \ln \left(\frac{\mu}{k_0}\right) - \ln \left(1 + \frac{1}{k_0}\right) \ldots - \ln \left(1 + \frac{z}{k_0}\right)$$

For the approximation of interest, k, k_0 and μ are large compared with unity so that the logarithm can be expanded in a power series and terms $[(\mu - k)/k_0]^2$ and higher powers neglected:

$$\ln P_\mu(k_0 + z) \simeq \ln P_\mu(k) \, e^{-z^2/2k_0} = C \, e^{-z^2/2k_0}$$

Substituting $z \approx k - \mu$ yields

$$P_\mu(k_0 + z) \simeq P_\mu(k_0) \, e^{-z^2/2k_0} = C \, e^{-z^2/2k_0} \tag{111}$$

It is convenient to make one further substitution:

$$h^2 = \frac{1}{2k_0} \tag{112}$$

The value of C may be determined by the normalization condition $C \int_{-\infty}^{\infty} e^{-h^2 z^2} dz = 1$, which yields (recalling that $\int_{-\infty}^{\infty} e^{-x^2} dx = \sqrt{\pi}$)

$$C = \frac{h}{\sqrt{\pi}}$$

Thus we conclude

$$P_\mu(k_0 + z) \simeq P_\mu(\mu + z) \simeq \frac{h}{\sqrt{\pi}} \, e^{-h^2 z^2} \tag{113}$$

This expression is the normal probability density function, or the normal differential probability distribution. Because it represents only one value in a continuum of values, it does not have the significance of probability until it is multiplied by a small interval Δz, or the differential dz. This differential normal probability distribution can be denoted

$$G_h\,(z)\,dz = \frac{h}{\sqrt{\pi}}\,e^{-h^2 z^2}\,dz \qquad (114)$$

and is the probability of observing a deviation z within the interval dz.

Three features of the normal probability distribution should be noted:

1. It is a continuous function of the variable $z = k\mu$ with limits $-\infty$ and $+\infty$.
2. It is symmetric about the mean μ (since z appears to second power only).
3. The parameter h is independent of the mean μ, and determines the maximum ordinate and the shape of the distribution.

The location and precision indexes previously defined can be readily obtained for the normal distribution.

Following the definition of Equation 110, the mean value can be evaluated:

$$z_{mean} = \frac{h}{\sqrt{\pi}}\,\int_{-\infty}^{\infty} z e^{-h^2 z^2}\,dz = 0 \qquad (115)$$

as expected. If the distribution were written in terms of the variable $x = z + u$ (x is a continuous representation of the original discrete variable k in Equation 110), then the mean value of x is

$$x_{mean} = \frac{h}{\sqrt{\pi}}\,\int_{-\infty}^{\infty} x e^{-h^2 (x-\mu)^2}\,dx = \mu \qquad (116)$$

Thus the mean value μ of the Poisson distribution does represent the mean value of the approximating normal distribution to within the accuracy of the approximation.

The *experimental standard deviation* s for a set of n measurements can be obtained directly by application of Equation 61, which is a general definition whether or not the data follow a normal distribution. The standard deviation provides, as does any dispersion index, an estimate of the likely range of values into which the next measurement may fall. With this interpretation, s is sometimes called the standard deviation *of a single measurement,* rather than of the distribution itself.

The parent standard deviation σ and variance σ^2 are of considerable importance. With z as the normal variable so that the mean is zero, then substitution into Equation 63 yields the variance as

$$\sigma^2 = \frac{h}{\sqrt{\pi}}\,\int_{-\infty}^{\infty} z^2 e^{-h^2 z^2}\,dz = -\frac{1}{h\sqrt{\pi}}\,\int_{-\infty}^{\infty} e^{-h^2 z^2}(-2h^2 z^2)\,dz$$

Integrating by parts yields

$$\sigma^2 = \frac{1}{2h^2} \quad \text{and} \quad \sigma = \frac{1}{h\sqrt{2}} \qquad (117)$$

The standard deviation in the mean, s_m or σ_m, can be obtained by application of Equations 73, 74, or 75.

The normal probability distribution is often expressed in terms of its mean and variance. By substitution

$$G_\sigma(x)dx = \frac{1}{\sigma\sqrt{2\pi}}\,e^{-(x-\mu)^2/2\sigma^2}\,dx \qquad (118)$$

The substitution $h^2 = \frac{1}{2} k_0$ involved (Equation 112) in our derivation of the normal distribution implies that $\sigma^2 = k_0$, and thus $\sigma = \sqrt{\mu}$. This is indeed the case for the normal distribution which approximates the Poisson distribution $P_\mu(k)$ with its standard deviation $\sigma = \sqrt{\mu}$. It must be remembered, however, that the parameter h and thus σ is a freely adjustable parameter for the normal distribution, and is not in general uniquely related to the mean value μ.

Figure 10 shows a plot of the differential normal probability distribution $G_o(x)dx$ with $\mu = 16$ and $\sigma = 4$. As evident, this is a good approximation to the Poisson distribution with $\mu = 16$ on the same figure.

The differential normal probability distribution $G_o(x)dx$ of Equation 118 is the probability of observing a deviation x within the interval dx. Thus the probability of observing a deviation x in the range between two particular values of x, say x_1 and x_2, is simply the integral of the normal probability distribution function $G_o(x)dx$ evaluated over the range between x_1 and x_2. This corresponds, of course, to the area under the normal density curve bounded by the two limits. For the purposes of standardization, let us write the normal density function $G_o(x)$ so that the mean $\mu = 0$ and x is expressed in units of σ ($\sigma = 1$). Then from Equation 118 we get

$$G_1(x) = \frac{1}{\sqrt{2\pi}} e^{-x^2/2} \tag{119}$$

Then the probability of finding a deviation x between the two limits x_1 and x_2 (where the x are in units of σ) is given by

$$\frac{1}{\sqrt{2\pi}} \int_{x_1}^{x_2} e^{-x^2/2} dx \tag{120}$$

Most often the limits of interest are symmetric about zero, so that $x_1 = -x_2 = -x'$ (say) and the integration is from $-x'$ to $+x'$. This particular integral is commonly listed in statistical tables[15] and available in computer subroutines. It is called the *error function*, abbreviated erf:

$$\text{erf}(x') = \frac{1}{\sqrt{2\pi}} \int_{-x'}^{x'} e^{-x^2/2} dx \tag{121}$$

Evidently, the integral value between arbitrary limits x_1 and x_2 can be deduced by simple additions or subtractions of the error function values.

If we are interested in the probability P_x, that a measurement of x will fall outside a certain range $-x'$ to x' (with x expressed in units of σ), then we can write

$$P_{x'} = 1 - \text{erf}(x') \tag{122}$$

since we know that the integral over the entire range $-\infty$ to $+\infty$ is equal to unity by normalization. The values of $P_{x'}$ are depicted graphically in Figure 11. For a normal frequency distribution, the probability of finding a deviation x greater than one standard deviation ($x' = 1$), either plus or minus, is seen from the graph to be about 0.32, and of finding x greater than 2 SD (plus or minus) to be about 0.046.

The *probable error* (P.E.) is a dispersion index defined so that the probability of finding deviations greater than plus or minus the P.E. equals 0.5. Thus for a normal distribution, it can be seen from Figure 11 that $P_{x'} = 0.5$ when $x' = 0.67$, so that the P.E. = 0.67σ.

As mentioned previously, the normal probability distribution occupies a central

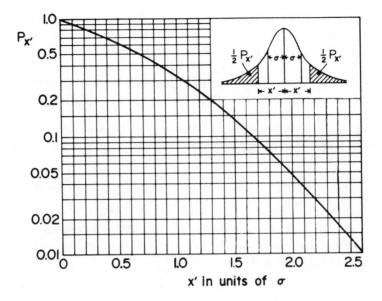

FIGURE 11. A graph of the probability P_x, that a measurement of x will fall outside a particular range from $-x'$ to $+x'$, assuming a normal distribution. This figure indicates the probability of various deviations, expressed in units of the standard deviation σ.

place in statistics and in the analysis of errors. The basic reason for this is the simplification involved if it can be assumed that the parent distribution is normal. Examples of such simplification are (1) the weight to be assigned to each measurement x_i with standard deviations S_{xi}, is $1/S_{xi}^2$ if the x_i are normally distributed; (2) the method of least squares is strictly valid if the normal distribution applies; and (3) errors propagate according to convenient rules if the parent distributions are normal. For such reasons (which are discussed further in Section II.C) it is often useful when dealing with statistical fluctuations in radioactive decay to approximate the relevant Poisson distribution by the corresponding normal distribution with standard deviation σ put equal to $\sqrt{\mu}$. This approximation is valid for large μ, and is often acceptable, particularly in the central region, for μ down to as small as ten (Figure 10).

4. The Interval Distribution

The three frequency distributions which have been dealt with here involved trials or events. In measurements involving radioactive decay, the time interval between events is often of importance. If the primary process is random, and follows Poisson's distribution, we can derive easily the distribution in size of the time intervals between successive events. Let the true mean rate have the constant value of a events per unit time. Then the probability that there will be no event in the time interval t, during which we would expect a events on the average, is, from Equation 97, $P_0(t) = e^{-at}$. The probability that there will be one event in the time dt is $P_1(dt) = adt$. Then the combined probability for no events in the time interval t and one event between t and t + dt is the product of these two independent probabilities ($P_0(t)$ and $P_1(dt)$). Therefore, the probability dP_t that the duration of a particular interval will be between t and t + dt is

$$dP_t = ae^{-at}dt \qquad (123)$$

if the random process is one which satisfies the conditions for Poisson's distribution.

(If the Poisson conditions are not satisfied, then the interval distribution will still be given by $dP_t = P_0(t) \cdot P_1(dt)$ but with $P_0(t)$ and $P_1(dt)$ given by the binomial distribution. Integration of Equation 123 from 0 to ∞ yields unity, verifying proper normalization.)

It is evident from Equation 123 that short intervals have a higher probability than long intervals for randomly distributed events, and in fact the most probable interval is zero.

Integration of Equation 123 from $t = t'$ to ∞ gives the probability that a particular interval will be longer than t', and hence the fraction of the intervals longer than t'. It is

$$P_{\geqslant t'} = \int_{t=t'}^{\infty} dP_t = e^{-at'} \tag{124}$$

where at' is the number of events expected, on the average, in the interval t'. Then the fraction of intervals that are shorter than t', which is equivalent to the probability that any particular interval be shorter than t', is

$$P_{\leqslant t'} = 1 - P_{\geqslant t'} = 1 - e^{-at'} \tag{125}$$

By definition of the mean rate as a events per unit time, it follows that the average interval is $1/a$. Then from Equation 124 the fraction of intervals larger than the average interval, $t' = 1/a$, is

$$P_{\geqslant 1/a} = e^{-1} \simeq 0.37 \tag{126}$$

The dispersion indexes for the interval distribution are easily obtained. In particular, the variance is given from Equation 63 as

$$\sigma^2 = \int_0^{\infty} \left(t - \frac{1}{a}\right)^2 a\, e^{-at} dt = \frac{1}{a^2} \tag{127}$$

and thus the standard deviation σ is equal to $1/a$, the average interval.

One may also obtain the distribution of intervals between random events counted in 2s, 3s, or more. Such an s-fold interval distribution shows an increasing suppression of the frequency of short intervals as s increases beyond one. A derivation of the s-fold interval distribution along with examples can be found in Evans.[12] Such distributions were of some importance when counting instruments were slow and prescalers were commonly used to regularize the interval distribution and reduce the counting losses. Now very fast electronic counters are readily available and prescaling circuits are seldom used.

5. Summary of Frequency Distributions

The basic formulas for the four frequency distributions which have been discussed are collected in Table 1, along with the expressions for their means and for their standard deviations. These are the most important location and dispersion indexes.

6. Some Numerical Examples

The initial task of a statistical treatment of data is usually to determine, by means of various tests, which frequency distribution provides an adequate model for the data. But for the analysis of data from a random process such as radioactive decay, this initial task can usually be accomplished very simply. The Poisson distribution and the interval distribution are generally applicable to such cases. And, except for those uncommon occasions when the number of events n is small, the normal distribution with

Table 1
FORMULAE OF THE FREQUENCY DISTRIBUTIONS DISCUSSED IN SECTION II.B, WITH THEIR MOST IMPORTANT LOCATION AND DISPERSION INDEXES

	Frequency distribution	Mean	Standard deviation
Binomial	$P_{n,p}(k) = \dfrac{n!}{k!(n-k)!}\, p^k q^{n-k}$	$\mu = np$	$\sqrt{npq}$
Poisson	$P_{\mu}(k) = \dfrac{\mu^k e^{-\mu}}{k!}$	μ	$\sqrt{\mu}$
Normal	$G_{\sigma}(x)dx = \dfrac{1}{\sigma\sqrt{2\pi}}\, e^{-(x-\mu)^2/2\sigma^2}\, dx$	μ	σ
Interval	$dP_t = ae^{-at}dt$	$1/a$	$1/a$

Table 2
DATA FROM TEN COUNTING TRIALS USING A SCINTILLATION DETECTOR AND SOURCE OF GAMMA ACTIVITY

(The sample mean and standard deviation are obtained, and the parent standard deviation estimated)

x_i	$\Delta x_i = x_i - m$	$(\Delta x_i)^2$
36,873	242	58,564
36,635	4	16
36,379	−252	63,504
36,935	304	92,416
36,614	−17	289
36,707	76	5,776
36,654	23	529
36,487	−144	20,736
36,405	−226	51,076
36,624	−7	49

$$m = 36,631 \qquad \Sigma(\Delta x_i)^2 = 292,955 \qquad s = \left(\frac{292,955}{10}\right)^{1/2} = 171 \qquad \sigma = \frac{10s}{9} = 180$$

an appropriate choice of standard deviation is also useful. The following examples should serve to illustrate various aspects from this Section II.B.

a. A Counting Experiment

A sample emitting gamma rays (from the annihilation of positrons following the decay of ^{22}Na) was repeatedly counted with a scintillation detector. The decay of the source ($T_{1/2}$ = 2.6 years) was negligible in the short period of the measurements (10 sec each, 5 min total). Table 2 lists the number of counts recorded in each of ten intervals. The mean m of these ten trial measurements x_i, as defined in Equation 53, is determined to be m = 36,631. Columns 2 and 3 in Table 2 list the deviations from the mean $\Delta x_i = x_i - m$ (Equation 57) and the deviations squared. This latter column is listed so that the standard deviation s can be evaluated (Equation 61) as $[\Sigma(\Delta x_i)^2/n]^{1/2}$, or in this case, 171. Our best estimate of the parent standard deviation σ can be expressed in terms of the experimental standard deviation s following Equation 64, or $\sigma = \sqrt{n/(n-1)}\,s = 180$.

It is reasonable to assume that the measured counts would be distributed according to the Poisson distribution about the parent mean $\mu \simeq m = 36,631$. Thus the standard deviation for a single measurement could be estimated from the relation of Equation 99, $\sigma = \sqrt{\mu} = 191$. This value is in reasonable agreement with the earlier evaluation of $\sigma = 180$ which made no assumption about the form of the frequency distribution.

The mean number of counts per interval has been evaluated as 36,631. The standard deviation of the mean is related to the standard deviation of a single measurement by Equation 74, $\sigma_m = \sigma/\sqrt{n}$. Thus we have $\sigma_m = 60$ and we can conclude that from our ten measurements we have determined the counting rate per 10 sec intervals to be $36,631 \pm 60$ with the indicated deviation being the standard deviation arising from the statistical fluctuations.

It is of interest to note that even though the measured distribution of events in this example does approximate a Poisson distribution (at least regarding its standard deviation), the measurements have not verified that the radioactive decay of the source satisfies the Poisson criteria. The reason for this is that there were other factors such as the solid angle subtended by the detector and the detector efficiency which themselves were subject to statistical fluctuations. This has been discussed in detail by Joyce.[16]

b. A Counting-Loss Problem

Suppose that the scintillation counting of the previous example was performed using a multichannel analyzer which had a 20 μsec dead-time, i.e., after each accepted event it took 20 μsec to process the pulse and blocked any subsequent events during this period. This is sometimes referred to as a nonparalyzable counter. For this type of counter, we can determine the fraction of the true counting rate which is lost due to the dead-time even without recourse to the interval distribution. If a and a' are the true and observed mean counting rates, and t' ($= 20$ μsec) is the dead-time, then the apparatus is dead a fraction of the time a't', and sensitive a fraction of time $1 - a't'$. This latter is thus the fraction of true events which can be recorded so that

$$\frac{a'}{a} = 1 - a't' \tag{128}$$

or

$$a = \frac{a'}{1 - a't'} \tag{129}$$

At relatively low counting rates such that $a't' \ll 1$ (small counting losses) Equation 129 yields the useful approximation:

$$a \approx a'(1 + a't') \tag{130}$$

Note also that the observed rate can be written in terms of the true rate as

$$a' = \frac{a}{1 + at'} \tag{131}$$

so that as a becomes very large (at $\gg 1$), a' approaches asymptotically the value $1/t'$. By then the apparatus is counting regularly — the statistical fluctuations of the source have been overwhelmed.

Substituting the particular values $a' = 3663$ sec^{-1} and $t' = 20$ μsec into Equation 128 yields the fraction of recorded events a'/a as 0.927. Thus the counting losses are 7.3%. When counting losses are appreciable, there will be some alteration of the frequency distribution of detected events because of the loss of short intervals. This is discussed in detail by Feller.[17]

Another type of apparatus will also lead to counting losses. This is the paralyzable type of detector, such as a Geiger counter, which stays dead for a time t' after the arrival of any event. Then only the pulses are counted with preceding intervals longer than t'. The interval distribution yields the fraction of intervals longer than t' to be (Equation 124) $e^{-at'}$, where a is the mean number of counts per unit time. This fraction is equal to the ratio of the observed to true counting rates a'/a.

For a sufficiently small that $at' \ll 1$, we can expand the exponential and write

$$a' \approx a(1 - at') \tag{132}$$

or

$$a \approx a'(1 + a't') \tag{133}$$

Note that Equation 133, the low counting loss approximation for paralyzable apparatus, is the same as Equation 130, the low counting loss approximation for nonparalyzable apparatus. Thus, providing counting losses are kept reasonably low, Equations 130 and 133 can be used to correct for losses without regard for the details of the apparatus.

Two sources of approximately equal activity can be used to estimate the dead time of a detector system by counting them first singly and then together.[18] If the counting rates are selected so that the counting losses are fairly low, then Equations 130 and 133 will be applicable and will provide a solution for t'.

c. Random-Coincidence Rate

Suppose that two Anger-type detectors are mounted on opposite sides of a patient and connected with a coincidence circuit to record the 511 keV gamma rays from annihilation of positrons emitted from a ^{18}F label (a positron camera[19]). Suppose also that the resolving time of each detector is $t' = 0.1$ μsec.

Detector one receives randomly distributed events at an average rate a_1, detector two receives events at a rate a_2, and the true coincidence rate is denoted $a_{1,2}$. Then the rates for single events not associated with any coincidence events are $(a_1 - a_{1,2})$ and $(a_2 - a_{1,2})$ for detectors one and two, respectively. The random-coincidence rate due to single events in detector one being followed, within its resolving time t', by random single events in detector two will be

$$(a_1 - a_{1,2}) \, t' \, (a_2 - a_{1,2})$$

There will be also a random-coincidence rate from single events in detector two being followed within its resolving time t' by random single events in detector one, given by

$$(a_2 - a_{1,2}) \, t' \, (a_1 - a_{1,2})$$

If $a_1 t'$ and $a_2 t'$ are much less than one, then the double random-coincidence rate will be negligible and the total coincidence rate is

$$a_{random} = 2t'(a_1 - a_{1,2})(a_2 - a_{1,2}) \tag{134}$$

Very often a_1 and a_2 will be much greater than $a_{1,2}$, in which case

$$a_{random} \approx 2t'a_1 a_2 \tag{135}$$

The rates a_1 and a_2 are usually proportional to the source activity I, so that the random coincidence rate is proportional to I^2. Because the true coincidence rate is proportional to I, as the source activity is increased the random coincidence rate will eventually dominate the true coincidence rate, imposing an upper limit on the useful source strength in a coincidence measurement.

For a positron camera with $t' = 0.1$ μsec, if $a_1 = a_2 = 10^5$/sec, then (from the approximation of Equation 135) $a_{random} = 2000$ counts/sec. The two annihilation gamma rays are correlated in direction so that if scattering and absorption of the gamma rays are negligible, then the true coincidence rate $a_{1,2}$ will be approximately given by the singles rate a_1 ($= a_2$) times the detector efficiency. Suppose this is 0.1, then the true coincidence rate $a_{1,2}$ in the hypothetical positron camera would be 10^4/sec. A better estimate of the random coincidence rate could now be obtained from Equation 134, yielding 1620 counts/sec.

C. Some Further Statistical Considerations

In this section we will deal very briefly with some topics which arise in the statistical treatment of data. For further details, readers are referred to one of the monographs on the subject.[9-11]

1. Propagation of Errors

Suppose f is a function of two *independently* observed quantities x and y, and that individual observations x_i and y_i have associated small deviations $\delta x_i = x_i - \bar{x}$ and $\delta y_i = y_i - \bar{y}$. Then function $f(x_i, y_i)$ can be expanded about the mean values $\bar{x}$ and $\bar{y}$ in a Taylor expansion and higher order terms neglected for small deviations. Thus, putting $u_i = f(x_i, y_i)$, we have

$$u_i = f[\bar{x} + \delta x_i), \ (\bar{y} + \delta y_i)]$$

$$= f(\bar{x}, \bar{y}) + \frac{\partial u}{\partial x} \delta x_i + \frac{\partial u}{\partial y} \delta y_i$$

and

$$\delta u_i = u_i - \bar{u} = \frac{\partial u}{\partial x} \delta x_i + \frac{\partial u}{\partial y} \delta y_i \tag{136}$$

where the partial derivatives are evaluated at $x = \bar{x}$ and $y = \bar{y}$. The mean $\bar{x}$ has been taken as $f(\bar{x}, \bar{y})$ here, but for small deviations this is equivalent to $\bar{u} = (\sum_{i=1}^{n} u_i)/n$.

The square of the standard deviation in u, written s_u, is given by definition and substitution from Equation 136 as:

$$s_u^2 = \frac{\sum_{i=1}^{n} (\delta u_i)^2}{n} = \frac{\left(\frac{\partial u}{\partial x}\right)^2 \Sigma(\delta x_i)^2 + 2 \frac{\partial u}{\partial x} \frac{\partial u}{\partial y} \Sigma(\delta x_i \delta y_i) + \left(\frac{\partial u}{\partial y}\right)^2 \Sigma(\delta y_i)^2}{n} \tag{137}$$

As n increases the sum $\Sigma(\delta x_i \, \delta y_i)$ goes to zero if the x_i and y_i are completely uncorrelated because positive and negative deviations will cancel. Since

$$s_x^2 = \frac{\Sigma(\delta x_i)^2}{n}$$

and

$$s_y^2 = \frac{\Sigma(\delta y_i)^2}{n}$$

then

$$s_u^2 = \left(\frac{\partial u}{\partial x}\right)^2 s_x^2 + \left(\frac{\partial u}{\partial y}\right)^2 s_y^2 \qquad (138)$$

This can be extended to more than two parameters in the obvious fashion; the general expression for J independent variables may be written:

$$s_u = \left[\sum_{j=1}^{J} \left(\frac{\partial u}{\partial x_j}\right)^2 s_{x_j}\right]^{1/2} \qquad (139)$$

Of frequent application are the formulas for the standard deviations of the sum, difference, product, and quotient of two quantities. These four cases can be represented by $u = x + y$, $x - y$, xy, and x/y. Then calculation of the partial derivatives $\partial u/\partial x$, $\partial u/\partial y$ and substitution into Equation 138 yields

$$s_{(x+y)} = (s_x^2 + s_y^2)^{1/2}$$

$$s_{(x-y)} = (s_x^2 + s_y^2)^{1/2}$$

$$s_{(xy)} = [xy(x + y)]^{1/2}$$

$$s_{(x/y)} = \left(\frac{x}{y^2} + \frac{x^2}{y^3}\right)^{1/2} \qquad (140)$$

The *standard deviation in the mean* was defined in Section II.A.2 and stated without proof in Equation 73. It can be treated now as an example of the compounding of errors. With reference to Equation 139, the standard deviation in the mean (or standard error, as it is sometimes called) can be written

$$s_m = \left[\sum_{i=1}^{n} \left(\frac{\partial \bar{x}}{\partial x_i}\right)^2 z_i^2\right]^{1/2} = \left[\frac{1}{n^2} \sum_{i=1}^{n} z_i^2\right]^{1/2} = \frac{s}{\sqrt{n}} \qquad (141)$$

in agreement with Equation 73. This expression for the standard deviation of the mean can be applied to any one of the independent x_i measurements in Equation 139, and from this it follows that

$$s_{\bar{u}} = \left[\sum_{j=1}^{J} \left(\frac{\partial u}{\partial x_j}\right)^2 s_{\bar{x}_j}^2\right]^{1/2} \qquad (142)$$

We note in passing that the distribution of mean values tends to be close to a normal distribution even if the parent distribution is not normal. Thus Equations 141 and 142 are usually applicable.

The discussion above pertains to random errors such as those arising from the fluctuations of radioactive decay. But a statistical assessment of the random errors in a result extracted from experimental data is of little or no benefit if there are large unrecognized systematic errors present which invalidate the estimated result. It is crucial to take all steps possible to eliminate systematic errors from measurement and analysis, and to estimate the magnitude of residual systematic errors which may be present.

2. Parameter Fitting — Method of Maximum Likelihood

It may be that the goal of a set of measurements such as a counting experiment is to fit the data to a known functional relation and to extract the best value for a certain parameter. As an example which we will treat in Section II.C.4, the decay constant of a radionuclide may be determined by fitting a set of activity data to the exponential decay law, thereby extracting the best possible estimate of λ.

Suppose that we wish to extract the most likely estimate γ_0 of a parameter γ in a known functional relation $\phi(x,\gamma)$. The method of maximum likelihood is generally considered the best method of estimation; as will be seen later, the method of least squares can be considered a special case of the method of maximum likelihood for normally distributed data.

Suppose that $\phi(x_i,\gamma)$ is a known (or assumed) functional relation representing the normalized probability density of getting a particular experimental result x_i involving the parameter γ. Then the likelihood function $L(x_i,\gamma)$ is the joint probability density of getting a particular experimental result x_1, x_2, ... x_n. Assuming the individual probabilities are uncorrelated, it will be the product of all n values of $\phi(x_i,\gamma)$:

$$L(x_1,x_2,\ldots x_n;\gamma) = \phi(x_1,\gamma)\phi(x_2,\gamma)\ldots\phi(x_n,\gamma) \tag{143}$$

If the experiment were repeated N times, the N different values of the likelihood function $L(x_i,\gamma)$ would themselves form a distribution. For large N these values of L approach a normal distribution with mean value at the maximum of the distribution corresponding to the best estimate of the parameter γ.[20]

The value of γ which makes L a maximum can be determined by differentiating L with respect to γ (assuming γ to be a continuous variable) and setting the derivative equal to zero. Because L is a maximum when its natural logarithm ln L is a maximum and because it is convenient to deal with a sum rather than with a product, we work with ln L. The most likely value γ_0 of the parameter γ is given by

$$\left(\frac{\partial \ln L}{\partial \gamma}\right)_{\gamma=\gamma_0} = 0 = \sum_{i=1}^{n}\left(\frac{\partial \ln \phi(x_i,\gamma)}{\partial \gamma}\right)_{\gamma=\gamma_0} \tag{144}$$

This procedure can be generalized for any number of parameters γ_i; in general there is one likelihood equation for each parameter.

A related concept is the likelihood ratio. For a particular set of experimental results x_i and parameter γ, the likelihood ratio R is the relative probability of two different values of the parameter, say γ_a and γ_b:

$$R = \frac{\displaystyle\prod_{i=1}^{n}\phi(x_i,\gamma_a)}{\displaystyle\prod_{i=1}^{n}\phi(x_i,\gamma_b)} = \frac{L(x_i,\gamma_a)}{L(x_i,\gamma_b)} \tag{145}$$

Obviously the likelihood ratio will be less than one if the parameter value γ_b in the denominator is the most likely value γ_0.

The above discussion of the maximum likelihood method assumes that there is no *a priori* knowledge of the parameter γ — all values are deemed equally probable before the experimental determination. If there is some known probability distribution for the parameter γ, say $H(\gamma)$, it can be included in the definition of the likelihood function (Equation 143) and the likelihood ratio (Equation 145).

As mentioned above, the distribution of values of the likelihood function approaches a normal distribution for large N. Thus in this limit we can express this distribution in terms of its variance σ^2 and mean γ_0 (Equation 118):

$$L(x_i, \gamma)\, d\gamma = C\, e^{-(\gamma-\gamma_0)^2/2\sigma^2}\, d\gamma$$

Differentiating twice leads to an expression for the variance of the distribution:

$$\frac{\partial \ln L}{\partial \gamma} = \frac{\partial}{\partial \gamma}\left[-\frac{(\gamma-\gamma_0)^2}{2\sigma^2} + \ln C \right] = -\frac{(\gamma-\gamma_0)}{\sigma^2}$$

and thus

$$\sigma^2 = -\left(\frac{\partial^2 \ln L}{\partial \gamma^2}\right)^{-1} \tag{146}$$

In a particular experiment which has measured x_i, an analytic evaluation of γ_0 and σ^2 using the above formulas may not be appropriate and a numerical approximation may be necessary. An application of the maximum likelihood method to obtaining the decay constant λ from a set of activity measurements is given as an example in Section II.C.4.

The relation of Equation 144 can be used to justify the method of least squares. Suppose that we have a single independent variable x and that we are concerned with values of y given by $y = f(x,\alpha)$ where α is a fixed parameter which we wish to determine. Suppose the deviations are all in y, so that $\delta y_i = y_i - f(x)$, and are normally distributed. Then for given values of α and x_i the probability of a measurement yielding a result between y_i and $y_i + dy_i$ is given by (Equation 66)

$$p(y_i, y_i + dy_i) = \frac{1}{\sqrt{2\pi}}\frac{dy_i}{\sigma_i}\, e^{-(\delta y_i)^2/2\sigma_i} \tag{147}$$

So the probability of observing y_i in the interval dy_i, y_2 in dy_2 .... and y_n in dy_n is the likelihood function

$$L = \prod_{i=1}^{n} \frac{1}{\sqrt{2\pi}}\frac{dy_i}{\sigma_i}\, e^{-(\delta y_i)^2/2\sigma_i} \tag{148}$$

if the deviations are independent.

Let us substitute $z_i = y_i/\sigma_i$ and consider z_i as the ith coordinate in an n-dimensional "error space". Then Equation 148 becomes

$$L = \frac{1}{(2\pi)^{n/2}}\, e^{-\frac{1}{2}\Sigma(\delta z_i)^2}\, dv \tag{149}$$

where dv is the elementary volume in error space. We see from Equation 149 that the relative likelihood of any set of errors or deviations in dv depends only on $\Sigma(\delta z_i)^2$, the square of the radius vector in error space. Thus the value of the parameter α which corresponds to the maximum value of L is that for which $\Sigma(\delta z_i)^2$, the sum of the squares of the deviations, is a minimum. This validates the principle of least squares for normally distributed errors. But it should be noted that it is the sum of the weighted squares of deviations, $\Sigma[(\delta y_i)^2/\sigma_i^2]$, that is to be minimized with the weighting factor $1/\sigma_i^2$. Thus in the determination of the mean value m of n measurements x_i, with different standard deviations σ_i, and with normally distributed errors, the weighted mean should be evaluated as given by

$$m^W = \frac{\sum\limits_{i=1}^{n} \frac{x_i}{\sigma_i^2}}{\sum\limits_{i=1}^{n} \frac{1}{\sigma_i^2}} \tag{150}$$

3. Tests for Goodness of Fit

Having obtained an estimate of some desired parameter from a set of experimental data, one frequently wants to know how well the data fitted the hypothesis used in extracting the parameter value. This is an extensive topic in the theory of errors in measurement; we shall only mention the most common methods and refer the reader to standard texts[9-11,20] for details.

a. The t-Test for Consistency of Means

This test provides a measure of the statistical consistency of two mean values drawn from two different sample sets. It can be used when the standard deviations of the means are not known. Suppose we have two sample sets, each of n measurements, with means $\bar{x}_1$ and $\bar{x}_2$. Then the t parameter is defined as

$$t = \frac{(\bar{x}_1 - \bar{x}_2)}{\sigma_{(x_1 + x_2)}} \sqrt{\tfrac{1}{2}n} \tag{151}$$

with $\sigma_{(x_1 + x_2)}$ denoting the standard deviation of the 2n data values. The value of t will fluctuate if different pairs of sample sets are considered, because the sample sets are of finite size. If an infinite number of pairs of sample sets are imagined, the corresponding values of t constitute a t distribution which can be expressed as (for a normal parent distribution)

$$f(t) = c \left(1 + \frac{t^2}{\nu}\right)^{-(\nu+1)/2} \tag{152}$$

where c is a normalization constant and ν is the number of degrees of freedom. Knowing the analytical form of t, or having access to tabulated values,[15] we can determine the probability that the value of t of a particular pair of sample sets will fall outside a specified range. Thus the consistency of two means falls under suspicion if their t value has a calculated probability of, say, less than 0.05.

b. The F-Test for Consistency of Standard Deviations

A probabilistic test for the consistency of the standard deviations of two sets of measurements can be arranged in a similar fashion to the previous t-test for consistency of the means: the assumption is made that both sample sets of measurements are from the same population, and then the probability of the validity of this assumption is tested.

The working parameter is now F, defined as

$$F = \frac{\sigma_{x_1}^2}{\sigma_{x_2}^2} = \frac{\frac{n_1}{n_1 - 1} s_{x_1}^2}{\frac{n_2}{n_2 - 1} s_{x_2}^2} \tag{153}$$

where s_{x_1} and s_{x_2} are sample standard deviations in the n_1 measurements of x_1 and in the n_2 measurements of x_2. The F values corresponding to an infinite number of pairs of sample sets constitute a (continuous) F distribution which has the analytic form, if the two sets of measurements are from the same normal parent distribution,

$$f(F) = cF^{(\nu_1 - 2)/2}(\nu_2 + \nu_1 F)^{-(\nu_1 + \nu_2)/2} \tag{154}$$

where c is the normalization constant and $\nu_1 = n_1 - 1$ and $\nu_2 = n_2 - 1$ are the respective numbers of degrees of freedom. As in the t test, we can calculate the probability that the value of F of the next sample pair of sets will fall outside a specified range. If it turns out that the experimental value of F deduced from Equation 153 is not very probable, then the consistency of the two standard deviations should be questioned.

c. The Chi-Square (χ^2) Test for Consistency of a Model Distribution

The Chi-square test was developed by Pearson[21] as a test for the overall goodness of fit of a model distribution to experimental data. The test parameter χ^2 is defined as

$$\chi^2 = \sum_i \frac{[(\text{observed value})_i - (\text{expected value})_i]^2}{(\text{expected value})_i} \tag{155}$$

where the summation is over n independent classifications into which the data have been grouped, with at least five events in each group (so that a normal approximation to the binomial distribution is valid). The "expected values" are those obtained from the assumed frequency distribution. The χ^2 values corresponding to an infinite number of trial sets of measurements constitute a continuous distribution

$$f(\chi^2)\, d(\chi^2) = \frac{(\chi^2)^{\frac{1}{2}\nu - 1}\, e^{-\frac{1}{2}\chi^2}}{2^{\frac{1}{2}\nu}(\frac{1}{2}\nu - 1)!}\, d(\chi^2) \tag{156}$$

where ν is the number of degrees of freedom (ν is usually $n - 1$, $n - 2$, or $n - 3$, depending on the number of constraints used in obtaining the model). This χ^2 distribution, $f(\chi^2)d(\chi^2)$ gives the differential probability that χ^2 will lie between χ^2 and $\chi^2 + d\chi^2$ as a function of χ^2 with ν as a parameter. Then the integral of this distribution from a particular value χ_c^2 to infinity equals the probability P that if the set of measurements were repeated the new value of χ^2 would be greater than χ_c^2. If χ_c^2 was the value from the first set, then P would equal the probability that the deviations observed in the second set would be greater.

$$P = \int_{\chi_c^2}^{\infty} f(\chi^2)d(\chi^2) \tag{157}$$

Statistical tables[15] list values of χ^2, ν and P, and Figure 12 provides a graphical representation of their relation.

An example of the use of the χ^2 parameter may clarify its usefulness. If the activity of a slowly decaying source is being measured by counting for 1 min periods, then a number n of observed values are recorded. The χ^2 test may be used to test the goodness of fit of the observed values to a simple exponential decay law. As defined in Equation 155,

$$\chi^2 = \sum_i \frac{(x_i - I(0)e^{-\lambda t_i})^2}{I(0)e^{-\lambda t_i}} \tag{158}$$

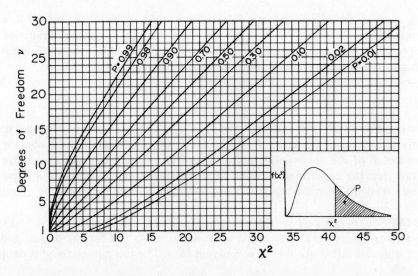

FIGURE 12. A graphical representation of the relation between the number of degrees of freedom, the value of the χ^2 parameter, and the probability P representing the integral of the $f(\chi^2)$ distribution from a particular value of χ^2 to infinity (the probability of seeing larger deviations in a second set of measurements).

Since λ is unknown there will be two constraints, one on the number of intervals and one on the intensity, so that ν will equal n − 2. Recourse to tables will yield P for this particular value of χ^2 and ν. It can then be decided if the model is "reasonable"; i.e., if P is less than some small value such as 0.05 the validity of the model becomes questionable.

It is noteworthy that if the expected value in the definition of χ^2 is a constant μ, and if the deviations follow a Poisson distribution, then from Equations 155, 61, and 102,

$$\chi^2 = \sum_{i=1}^{n} \frac{(x_i - \mu)^2}{\mu} = \frac{ns^2}{\sigma^2} \tag{159}$$

In this case, χ^2 is n times the ratio of the experimental variance to the parent variance.

It has been assumed in this example that the decay of the source is negligible during each counting period so that the conditions of validity for the Poisson distribution are upheld. The statistics of a rapidly decaying source have been treated by Peierls.[22]

4. Some Numerical Examples
This section will provide a few examples to illustrate the applicability of the relations outlined above for the propagation of errors, for the tests for goodness of fit, and for the least squares and maximum likelihood methods.

a. The Contribution of Background in a Counting Measurement
This example will be a simple yet important case illustrating the propagation of errors. Suppose scintillation counting is being performed on a series of samples. For a particular sample, 24,140 counts were recorded by the detector in 20 sec. The background was evaluated by counting for 1 min and 8617 events were recorded. The total activity measured for that particular sample was 1207 counts/sec, of which 144 counts/sec were background, leaving a net activity of 1063 counts/sec. The standard deviations σ for the total counts and the background counts are $\sqrt{24,140}$ and $\sqrt{1207}$, respectively, so that standard deviations of the total activity detected per second and of

the background per second are 8.8 and 1.5/sec. The standard deviation in the net activity can now be obtained by application of the second part of Equation 140, $\sigma_{(x-y)} = (\sigma_x^2 + \sigma_y^2)^{1/2}$ which determines the standard deviation for the net activity of 1063 counts/sec to be 9 counts/sec.

In this example the counting times for the sample and background measurements were arbitrarily selected. If it is desired to make the most efficient use of total time T, how should the counting times t_1 for the sample and t_2 for the background be divided? Let the sample and background counts be n_1 and n_2 so that the measured activities are $I_1 = n_1/t_1$ and $I_2 = n_2/t_2$. The net activity is $I = I_1 - I_2$. Then

$$\sigma^2(I) = (I_1) + (I_2) = \frac{n_1}{t_1^2} + \frac{n_2}{t_2^2} = \frac{I_1}{t_1} + \frac{I_2}{T - t_1} \tag{160}$$

σ^2 (and hence σ) can be minimized with respect to t_1 by differentiating the final expression of Equation 160 with respect to t_1, and setting the result equal to zero. One obtains

$$\frac{t_1}{t_2} = \sqrt{\frac{I_1}{I_2}} \tag{161}$$

In the scintillation counting example above, the ratio $\sqrt{I_1/I_2}$ is about 3/1 so that a better proportioning of the counting time (considering one sample only) would have been 1 min counting the sample and 20 sec counting the background.

b. Consistency of Two Means

The counting measurements described in Section II.B.6 (Table 2) were repeated and a new mean value for ten measurements was determined as 36,764 ± 61. Is the new mean consistent with the old? Because we know the standard deviation of the two means, one possibility is to compare the difference of these two mean values, 133 counts, with the statistical standard deviation of the difference. With reference to Equation 140 the standard deviation of the difference of the two mean values can be determined to be 86. The measured difference of 133 falls at about 1.5 SD. Assuming a normal distribution to be a reasonable approximation for the distribution of mean values, Figure 11 indicates that a deviation of 1.5 SD or greater has a probability of 14%.

Another approach which makes use of the standard deviation of the measured data rather than of the mean values, is to apply the t-test of Equation 151, determining a t value of 1.7. Reference to tables[15] indicates that for 18 degrees of freedom, this value of t is exceeded about 12% of the time. Both of these tests of consistency suggest that these two mean values are not necessarily inconsistent; however, they are sufficiently separated that consideration should be given to the possibility of systematic variations.

c. A Chi-Square Test

The data in Table 2 can be subjected to a chi-square analysis as a test of the statistical consistency. The model we assume is simply a constant activity, and the ten measurements in the table can serve as ten independent groupings of data. Then the "expected value" is the mean, m = 36,631, and with reference to Equation 155, and Table 2, the value of the test parameter χ^2 is $\Sigma(\Delta x_i)^2/m = 8.0$. Referring now to Figure 12, it can be determined that for $\chi^2 = 8$ with 9 degrees of freedom (the use of the mean value imposes one constraint) the probability $P = 0.5$. This means that for data normally distributed about the mean m, there would be a 50% chance of obtaining larger deviations (a larger value of χ^2) on repeating the series of measurements. Thus the data of Table 2 do conform to the assumption of a normal distribution (or a Poisson distribution, since m is large) about a constant mean value. This conclusion may have been drawn earlier but less convincingly from the close agreement between the sample standard deviation s and the parent Poisson standard deviation σ.

d. Determination of a Decay Constant by Least Squares and Maximum Likelihood Methods

Example a in Section I.A.5 has shown a graphical approach to the determination of the decay constant λ from a set of activity measurements on a radioactive sample. This example shows a numerical method which is useful when higher accuracy or an indication of the statistical standard deviation of the result is required. The numerical method can be extended to a more complicated model than a single exponential decay. We shall assume that only one radioactive species is present and that the decay law is a simple exponential. We divide our total counting time into equal time intervals, and record (e.g., with a scintillation counter) the number of observed decay events in each time interval. The multiscaling mode of a multichannel analyzer provides a convenient way to perform such a measurement. The total counting time is selected so that it encompasses at least one half-life of the radioactive species.

The number of disintegrations of a radioactive species in the time interval between t_i and $t_i + \Delta t$ can be obtained by integrating Equation 6 over that time interval:

$$\int_{t_i}^{t_i+\Delta t} I(0)\, e^{-\lambda t} = \frac{-I(0)}{\lambda}\, e^{-\lambda t} \Bigg]_{t_i}^{t_i+\Delta t} = \frac{I(0)}{\lambda}\, (1 - e^{-\lambda \Delta t})\, e^{-\lambda t_i} \qquad (162)$$

For time intervals Δt much smaller than the mean lifetime ($\lambda \Delta t \ll 1$), Equation 162 reduces to approximately $I(0)\,\Delta t\, e^{-\lambda t_i}$.

The expected number of counts μ_i in each time channel t_i will depend on the detector geometry and efficiency, but will be proportional to the above number of disintegrations in the corresponding interval:

$$\mu_i = K e^{-\lambda t_i} \qquad (163)$$

The actual number of counts y_i recorded in each time channel will be related to the mean μ_i through the Poisson probability distribution (Equation 96):

$$P_{\mu_i}(y_i) = \frac{\mu_i^{y_i}}{y_i!}\, e^{-\mu_i} \qquad (164)$$

If the number of expected counts μ_i in each time channel is not too small (say > 10) then we can approximate the expected Poisson probability distribution by a normal probability distribution (Equation 118),

$$G_{\mu_i}(y_i) = \frac{e^{-(y_i-\mu_i)/2\sigma_i^2}}{\sigma_i\, 2\sqrt{\pi}} \qquad (165)$$

having variance σ_i^2 equal to μ_i.

There are two methods for extracting the most likely value for the decay constant λ from our hypothetical set of decay measurements y_i. The most common is to assume that the normal distribution provides a good approximation and to use the method of least squares, which we have seen in Section II.C.2 to be valid for normally distributed errors. The independent variable is the time, which we assume to be free of errors.

The squared deviations are $(y_i - \mu_i)^2$, with the parent means μ_i being obtained by substitution from Equation 163. In this case the variances $\sigma_i^2 = \mu_i$ are different for each value of i so that we must minimize the sum of the *weighted* deviations. There are two unknown parameters λ and K and we are looking for their values which minimize the weighted sum of the squared deviations. Thus we obtain the two equations

$$\frac{\partial}{\partial \lambda} \sum \frac{(y_i - Ke^{-\lambda t_i})^2}{\sigma_i^2} = 0$$

and

$$\frac{\partial}{\partial K} \sum \frac{(y_i - Ke^{-\lambda t_i})^2}{\sigma_i^2} = 0 \tag{166}$$

with the summation over all the time channels, say i = 1 to N. A customary simplification is to approximate the parent mean $\mu_i = \sigma_i^2$ involved in the weighting factor by the experimental value y_i for each particular time channel. Then the two Equations 166 yield:

$$\sum t_i = \sum \frac{Ke^{-\lambda t_i}}{y_i}$$

and

$$\sum e^{-\lambda t_i} = \sum \frac{Ke^{-2\lambda t_i}}{y_i} \tag{167}$$

Equations 167 are nonlinear in the parameter λ. The problem becomes linear if the logarithms of the data and of Equation 163 are considered. Denoting ln y_i by y_i' and ln K by k', the squared deviations become $(y_i' - K' + \lambda t_i)^2$. The appropriate variance to be used in the weighting factor may be obtained using Equation 138 for the propagation of errors; one obtains $\sigma_{y_i}^2 = \sigma_{y_i}^2/y_i^2$. Now the weighted sum to be minimized is $\sum y_i(y_i' - K' + \lambda t_i)^2$. This is now a linear problem, equivalent to finding the best weighted fit of a straight line $y = mx + b$ to a set of data. It is also equivalent to the linear regression problem. Thus one can obtain a value for the best estimate of λ and of its standard deviation s_λ by substituting into standard formulae.[9-11] We can easily produce the relevant formulae. Differentiating the above expression for the weighted sum of deviations with respect to both parameters λ and K' and setting equal to zero yields the two equations

$$K'\sum y_i - \lambda \sum y_i t_i = \sum y_i y_i'$$

and

$$K'\sum y_i t_i - \lambda \sum y_i t_i^2 = \sum y_i y_i' t_i \tag{168}$$

These can be solved for λ using determinants:

$$\lambda = \frac{-\begin{vmatrix} \Sigma y_i & \Sigma y_i y_i' \\ \Sigma y_i t_i & \Sigma y_i y_i' t_i \end{vmatrix}}{\begin{vmatrix} \Sigma y_i & -\Sigma y_i t_i \\ \Sigma y_i t_i & -\Sigma y_i t_i^2 \end{vmatrix}} = \frac{\Sigma y_i \Sigma y_i y_i' t_i - \Sigma y_i y_i' \Sigma y_i t_i}{\Sigma y_i \Sigma y_i t_i^2 - (\Sigma y_i t_i)^2} \tag{169}$$

The standard deviation s_λ of this value of λ can be considered as propagating from the individual deviations of the y_i. Again invoking Equation 138,

$$s_\lambda^2 = \Sigma \left(\frac{\partial \lambda}{\partial y_i} \right)^2 s_{y_i}^2 \tag{170}$$

The expression for λ can be substituted from Equation 169 and using $s_{y_i}^2 \approx y_i$ yields an expression for s_λ.

The above example represents the simplest case of fitting a decay constant to a set of activity measurements. Additional complications may be present. For example, there may be a background counting rate. If it is constant in time, the easiest way to treat a background is to determine independently a best value for it, and then to subtract it from all measurements. The above analysis is then still applicable, except that the weighting factor must be appropriately modified. More than one radioactive species may be present so that a sum of more than one exponential decay term must be fitted to a set of data. Then the problem cannot be reduced to a linear one by taking logarithms; the appropriate equations must be solved iteratively. Parameter values obtained from a least squares analysis involving more than one exponential decay term should be viewed with some caution as the exponential decay constants are very strongly correlated.

A more general method of fitting a functional relationship to some experimental data than the previously discussed method of least squares is the method of maximum likelihood. The likelihood function $L(y_i, \lambda)$ is constructed from the product of the individual probability distributions for the values of the data points y_i about the parent means μ_i (Equation 143):

$$L = \prod_{i=1}^{n} \phi(y_i, \lambda) \tag{171}$$

with the $\phi(y_i, \lambda)$ given, for the two cases of Poisson and normal distributions, respectively, by (Equations 164 and 165):

$$\phi(y_i, \lambda) = \frac{\mu_i^{y_i} e^{-\mu_i}}{y_i!}$$

and

$$= \frac{e^{-(y_i - \mu_i)^2 / 2\sigma_i^2}}{\sigma_i 2\sqrt{\pi}} \tag{172}$$

We can substitute as before $\mu_i = K e^{-\lambda t_i}$ and $\sigma_i^2 = \mu_i \simeq y_i$.

As described in Section II.C.2, we obtain the most probable values of λ and K by determining those values which maximize the likelihood function; i.e., we take the derivatives of ln L with respect to λ and K and set equal to zero. For a Poisson distribution

$$\frac{\partial \ln L}{\partial \lambda} = \Sigma \frac{\partial}{\partial \lambda} \{ y_i \ln(Ke^{-\lambda t}i) - \mu_i - \ln(y_i!) \} = 0 \qquad (173)$$

and similarly for $\partial / \partial K$.

For the normal distribution

$$\frac{\partial \ln L}{\partial \lambda} = \Sigma \frac{\partial}{\partial \lambda} \left\{ \frac{(y_i - Ke^{-\lambda t}i)^2}{2y_i} + \ln(2\sqrt{\pi y_i}) \right\} = 0 \qquad (174)$$

and similarly for $\partial / \partial K$. In the case of the normal distribution (Equation 174) we obtain the equations

$$\Sigma t_i = \Sigma \frac{K}{y_i} e^{-\lambda t}i$$

and

$$\Sigma e^{-\lambda t}i = \Sigma \frac{K}{y_i} e^{-2\lambda t}i \qquad (175)$$

in agreement with those previously obtained by the method of least squares (Equation 167). The above expressions for λ must be solved iteratively.

If the shape of $L(y_i,\lambda)$ is approximately Gaussian in the vicinity of its maximum (and this should be numerically verified), then Equation 146 can be used to determine the standard deviation of λ:

$$s_\lambda = \left(\frac{-\partial^2 \ln L}{\partial \lambda^2} \right)^{-1/2} \qquad (176)$$

Detailed application of the maximum likelihood method to fitting the sum of exponential decay components can be found in two references where the method is applied to analysis of short decay lifetimes[23] and of tracer kinetic data.[24]

In fitting a functional expression to experimental data we have started with some assumptions about the functional form (e.g., a single exponential decay with a normal probability distribution). Once the best estimate and standard deviation have been obtained for the parameter of interest, it is prudent to perform a chi-square analysis as a test of the model assumptions. A calculation of the χ^2 parameter can be incorporated into any calculator or computer routine which evaluates the fitted parameter values. If the value obtained for χ^2 yields a very small probability P (Figure 12) for the appropriate number of degrees of freedom, then the model assumption should be viewed with suspicion. On the other hand, if a value of P larger than 0.5 or even 0.1 is obtained, the assumptions are not verified but may be considered reasonable within the limits of the statistical precision of the measurements.

REFERENCES

1. Rutherford, E. and Soddy, F., *Philos. Mag.,* 4, 370, 1902; 4, 569, 1902; 5, 576, 1903.
2. Hahn, H. P. et al., Survey on the rate perturbation of nuclear decay, *Radiochem. Acta,* 23, 23, 1976.
3. Ruark, A. E., *Phys. Rev.,* 44, 654, 1933.

4. von Schweidler, E., Premier Congres International de Radiologie, Liege, 1905.
5. Winter, R. G., Evolution of a quasi-stationary state, *Phys. Rev.,* 123, 1503, 1961.
6. Bateman, H., Solution of a system of differential equations occurring in the theory of radioactive transformations, *Proc. Cambridge Philos. Soc.,* 15, 423, 1910.
7. Wieland, B. W. and Highfill, R. R., *IEEE Trans. Nucl. Sci.,* NS-26, 1713, 1979.
8. Porter, W. C., Dworkin, H. D., and Gutkowski, R. F., *J. Nucl. Med.,* 17, 704, 1976.
9. Parratt, L. G., *Probability and Experimental Errors in Science,* John Wiley & Sons, New York, 1961.
10. Young, H. D., *Statistical Treatment of Experimental Data,* McGraw-Hill, New York, 1962.
11. Stevenson, P. C., *Processing of Counting Data,* National Academy of Sciences, Washington, D.C., 1966.
12. Evans, R. D., *The Atomic Nucleus,* McGraw-Hill, New York, 1955, chap. 26-28.
13. Bacon, R. H., *Am. J. Phys.,* 21, 428, 1953.
14. Bateman, H., *Philos. Mag.,* 20, 704, 1910.
15. Beyer, W. H., Ed., *Handbook of Tables for Probability and Statistics,* 2nd ed., Chemical Rubber Co., Cleveland, 1968.
16. Joyce, W. B., *Am. J. Phys.,* 37, 489, 1969.
17. Feller, W., *Studies and Essays,* Interscience, New York, 1948.
18. Beers, Y., *Rev. Sci. Inst.,* 13, 72, 1942.
19. Muehllehner, G., *J. Nucl. Med.,* 16, 653, 1975.
20. Fisher, R. A., *Statistical Methods for Research Workers,* 11th ed., Oliver and Boyd, Edinburgh, 1950.
21. Pearson, K., *Philos. Mag.,* 50, 157, 1900.
22. Peierls, R., *Proc. Roy. Soc. London,* A149, 467, 1935.
23. Orth, P. H. R., Falk, W. R., and Jones, G., *Nucl. Inst. Meth.,* 65, 301, 1968.
24. Sandor, T., Conroy, M. F., and Hollenberg, N. K., *Math. Biosci.,* 9, 149, 1970.

INDEX

A

Absorption coefficients, 89—94
Absorption of photon energy, 88
Activity, 38
    partial, 109
    ratio of between parent and daughter, 114—117
Addition rules of angular momentum, 5
Alpha decay, 48—59
    energetics of, 48—51
    theory of, 52—59
Alpha particles, 38
    spectra of, 51—52
Angular momentum, 4—5
    addition rules of, 5
    doubly magic nuclides, 19
    neighboring nuclides, 19
    orbital, 4, 6, 14, 16, 18
    permanently deformed nucleus and, 26
Annihilation radiation, 42, 80
Antineutrino, 41
Arithmetic average, see Mean
Atomic number, 39
Attenuation coefficients, 89—94
Average, see Mean
Average lifetime, 109
Axis, 24

B

Background in counting experiment, 154—155
Bartlett exchange, 11, 13
Base state, 28
Bateman equations, 112
Beta decay, 41—43
    analysis of, 59—68
    energetics of, 60—63
    Fermi theory of, 63—67
Beta particles, 38, 41
    spectra of, 59—60
Binding energy, 3—4, 9, 47—48
Binomial distribution
    frequency, 132—135
    probabilities, 133
Bohr stopping power formula, 96
Bragg-Kleeman rule, 99
Bremsstrahlung, 80, 101

C

Carbon-14, 122—124
Central forces, 9, 14
Characteristic X-rays, 80
Charge
    conservation of, 40
    zero, 63
Charged particles, 80

    heavy, 94—100
Charge independent forces, 13
Charge symmetric forces, 13
Chart of nuclides, 40
Chi-square test, 153—156
Closed shell hole, 13
Coherent Compton scattering, 88
Collective models, 22
Collective motion, 8, 22
Collision cross section, 81
Compton effect, 84—88
Compton scattering, 92
    coherent, 88
Conservation, 8, 40, 41
    of electric charge, 40
    of leptons, 41, 42
    of linear momentum, 42
    of momentum, 42, 49
    of nucleons, 40
    of parity, 66
Consistency
    of means, 152, 155
    of model distribution, 153—154
    of standard deviation, 152—153
    tests for, 152—153
Continuum theory, 107—126
Conversion, 46—47, 76—77
Core
    inert, 5, 8, 13, 22, 26
    repulsive, 9, 11
Coulomb repulsion, 3
Counting experiment, 145—146
    background in, 154—155
Counting-loss problem, 146—147
Coupling, 23, 25
Cross section
    collision, 81
    differential, 82, 84, 85
    electronic, 81
    pair production, 89
    photoelectric, 84
    total, 86
Curie (unit), 38

D

Daughters, 114—117
DeBroglie relationship, 80
Decay
    alpha, see Alpha decay
    beta, see Beta decay
    carbon-14, 122—124
    constant, see Decay constant
    continuum theory of, 107—126
    curve of, 38—39
    exponential, 107—108
    fundamental law of, 107
    gamma, see Gamma decay
    iodine-131, 124—125
    positron, 42

radioactive, 38—39, 119—121
random nature of, 106
statistical fluctuations in, 126—159
systematics of, 39—47
Decay constant, 39, 107—108, 156—159
daughter's and parent's nearly equal, 115
partial, 109—110
Decoupling constant, 28
Deformation
oblate, 24, 26
parameter of, 31
prolate, 24, 26, 30
Deformed nucleus, 3, 14, 22, 26, 27, 29—34
permanently, see Permanently deformed
nucleus
rotation of, 22, 27
vibration of, 22, 23
Delta rays, 94
Density of nucleus, 2—3, 9
Deuteron, 8, 9, 11—13
Deviation
mean (average), 129
standard, see Standard deviation
Differential cross section, 82, 84, 85
Di-neutron, 8, 13
Dipole
electric, 71
magnetic, 6, 71
vibrations of, 24
Di-proton, 8, 13
Disintegration, 38, 39
Dispersion indexes, 128—132
Distorted nucleus, 22
Distribution, 153—154
Doubly magic nuclides, 4, 13—22
angular momentum of, 19
energy states for, 15
even-even nuclides far removed from, 28, 34
excited states of, 19
odd-even, 19
odd-even nuclides far removed from, 31
odd-odd, 19
odd-odd nuclides far removed from, 31—34
parity of, 19
quantum numbers for, 18
spin of, 19
Drop model, 3

E

Elastic scattering, 102—103
Electric charge conservation, 40
Electric dipole, 71
Electric moments, 7—8
Electric monopole, 71
Electric multipole radiation, 72, 74
Electric quadrupole, 2, 7, 12, 26, 71
Electromagnetism, 8, 80
Electron capture, 41, 42, 62
experimental observations of, 68
ratio of to positron emission, 67—68

Electronic cross section, 81
Electron neutrino, 41
Electrons, 80, 100
interaction between matter and, 100—101
negative, 41
Emission
gamma rays, 68—76
positrons, 62, 67—68
End products, 121—122
Energetics
alpha decay, 48—51
beta decay, 60—63
Energy
binding, 3—4, 9, 47—48
doubly magic nuclides, 15
half-life and, 52
highly relativistic, 100
neighboring nuclides, 15
pairing, 4, 19, 22, 27
photon, 88
rotating permanently deformed nucleus, 27
symbolism for, 16, 18, 31
vibrating permanently deformed nucleus, 29
Equilibrium, 117
Error propagation, 148—149
Even-even nucleus, 27, 28
Even-even nuclides, 24, 34
Exchange
Bartlett, 11
force of, 9—10, 13
Heisenberg, 11
Majorana, 10
spin, 10
Excitations
doubly magic nuclides, 19
neighboring nuclides, 19
nucleus, 43
phonon-octupole, 29
rotational, see Rotational excitations
single particle, 16, 31
vibrational, 29
Exclusion principle, 11, 13, 14, 16
Experimental observations of electron capture, 68
Exponential decay, 107—108
Extrapolated range, 97

F

Fermi theory of beta decay, 63—67
Field theory, 8, 13
Fit goodness, 152—154
Fluctuations in radioactive decay, 126—159
Forces
Bartlett, 11, 13
central, 9, 14
charge independent, 13
charge symmetric, 13
electromagnetic, 8, 80
exchange, 9—10, 13
Heisenberg, 13
Majorana, 10, 13

noncentral, 12
nuclear, see Nuclear forces
pairing, 4, 19, 22, 27
tensor, 12
two-nucleon, 8—13
velocity-dependent, 12
Wigner, 9, 13
Fractional standard deviation, 132
Frequency distributions, 127—132
binomial, 132—135
interval, 143—144
mathematical models of, 132—148
moments of, 130
normal, 139—143
Poisson, 135—139
Frequency of vibration, 14
F-test for consistency of standard deviation,
152—153

G

Gallagher-Moszkowski rule, 34
Gamma decay, 43—46
analysis of, 68—77
Gamma rays, 38, 43
spectra of, 68
theory of emission of, 68—76
X-ray interactions with, 80—94
Generators, 125—126
Global properties of nucleus, 2—8
Goodness-of-fit tests, 152—154
Graphical solutions, 122
Ground state, 2, 44
Growth
of daughter, 112—114
radioactive, 119—121

H

Half-life, 39, 74, 108—109
energy and, 52
Half value layer, 83, 90
Harmonic oscillator, 16, 18, 24, 29
simple, 14
Harmonics, 23
Heavy charged particles interaction with matter,
94—100
Heavy lepton, 41
Heisenberg exchange, 11, 13
Highly relativistic energies, 100
Hole in closed shell, 13

I

Impact parameter, 94
Impurities in iodine-123, 125
Independent motion of nucleons, 13
Inelastic scattering, 103
Inert core of nucleus, 5, 8, 13, 22, 26

Internal conversion, 46—47
coefficients of, 76—77
Interval frequency distribution, 143—144
Intrinsic spin, 4, 16
Iodine-123 impurities, 125
Iodine-131 decay, 124—125
Ionization potential, 97
Isobars, 49
Isomers, 49
Isospin, 13
Isotones, 49
Isotopes, see Nuclides

K

Klein-Nishina formula, 85

L

Least squares, 156—159
Leptons, 41, 42
Life, 108—109
Linear momentum conservation, 42
Liquid drop model, 3
Location indexes, 127—128
Logarithmic decrement, 103

M

Magic nuclides, 3, 4, 13, 15, 16, 18, 28, 31—34
nuclides far removed from, 26—34
Magnetic dipole, 6, 71
Magnetic moments, 6—7
Magnetic multipoles, 71
radiation from, 72, 74
Magnetic quadrupole, 71
Majorana exchange, 10, 13
Mass absorption coefficient, 91
Mass at rest, 41
Mass number, 39
Mass stopping powers, 100
Mathematical models of frequency distribution,
132—148
Matrix element, 64
Matter interactions, 94—101
Maximum likelihood method, 150—152,
156—159
Mean, 127, 128
consistency of two, 155
t-test for consistency, 152
Mean deviation, 129
Mean ionization potential, 97
Mean life, 108—109
Mean path length, 90
Mean range, 97
Median, 127
Meson field theory, 8, 13
Mesons, 8
Mirror nucleus, 13

Mixture stopping power, 97
Mode, 127
Models, see also specific models
  chi-square test for consistency of distribution
        of, 153—154
  collective, 22
  frequency distribution, 132—148
  liquid drop, 3
  Nilsson, 26, 31
  nuclear, 2
  shell, see Shell
  single particle, 14, 16, 18
  vibrational, 22, 23
Molybdenum-99 generator, 125—126
Moment
  electric quadrupole, 2, 7, 12
  magnetic dipole, 6
Momentum
  angular, see Angular momentum
  conservation of, 49
  linear, 42
Monopole, 71
Motion
  collective, 8, 22
  rotational, 22, 27
  vibrational, 22
Multipole radiation
  electric, 72, 74
  magnetic, 72, 74
Multipoles
  electric, 72, 74
  magnetic, 71
Muon neutrino, 41

N

Natural radioactivity, 38
Negative electron, 41
Neighboring nuclides
  angular momentum of, 19
  energy states for, 15
  excited states of, 19
  odd-even, 19
  odd-odd, 19
  parity of, 19
  quantum numbers for, 18
  spin of, 19
Neutrinos
  electron, 41
  muon, 41
  properties of, 63
  tau, 41
Neutron number, 39
Neutrons, 39, 80
  capture of, 103
  elastic scattering of, 102
  interactions of matter with, 102—104
  symmetry of, 4, 11
Nilsson model, 26, 31
Noncentral force, 12
Nordheim's rules, 19, 31

Normal frequency distribution, 139—143
Nuclear density saturation, 2
Nuclear forces, 8, 13
  range of, 9, 11, 12
  saturation of, 9
Nuclear models, 2
Nuclear reactions, 103
  production of radionuclides by, 117—119
Nuclear states, 13—34
Nucleon number, 39
Nucleons, 39
  conservation of, 40
  independent motion of, 13
  pairing of, 4—6, 19
Nucleus
  deformed, see Deformed nucleus
  density of, 2—3, 9
  disintegration of, 39
  distorted, 22
  doubly magic, see Doubly magic nuclides
  energy states of, 31
  even-even, 27, 28
  excited states of, 43
  global properties of, 2—8
  inert core of, 5, 8, 13, 22, 26
  mirror, 13
  models of, 14
  neighboring, 18
  odd-even, 8, 27
  oscillations of, 23
  permanently deformed, 26, 27, 29—34
  rotating permanently deformed, 27
  rotational motion of, 22, 27
  shape of, 2—3, 13
  shell model of, 14
  single particle model of, 14, 16, 18
  spherical, 13, 15, 18
  spheroidal, 26
  spin of, 4, 18
  symbolism for energy states of, 16, 18, 31
  vibrational motion of, 22
  volume of, 9
Nuclides, 39, 49
  chart of, 40
  doubly magic, see Doubly magic nuclides
  even-even, 24, 34
  far removed from magic numbers, 26—34
  neighboring, see Neighboring nuclides
  odd-even, 5, 25, 31
  odd-odd, 31—34
  production of by nuclear reactions, 117—119
Number-distance curve, 97, 101

O

Oblate deformation, 24, 26
Octupole vibrations, 24
Odd-even nucleus, 8, 27
Odd-even nuclides, 5, 25
  doubly magic, 19
  far removed from numbers, 31

neighboring, 19
Odd-odd nuclides
  doubly magic, 19
  far removed from doubly magic numbers, 31—34
  neighboring, 19
Orbital angular momentum, 4, 6, 14, 16, 18
Oscillations of nucleus, 23
Oscillators, harmonic, see Harmonic oscillators

P

Pairing, 4—6, 19
  forces of, 4, 19, 22, 27
Pair production, 88—89, 92
  cross section of, 89
Parameter fitting, 150—152
Parents, 114—117
Parity, 5—7, 10, 16, 18, 72
  conservation of, 66
  doubly magic nuclides, 19
  neighboring nuclides, 19
  single particle excitations in permanently deformed nucleus, 31
  vibrational excitations, 29
Partial activity, 109
Partial decay constants, 109—110
Particles
  alpha, 38, 51—52
  beta, 38, 41, 59—60
  charged, 80
  excitations of, 16
  heavy charged, 94—100
  single, 14, 16, 18
  uncharged, 80
Path length, 90
Pauli exclusion principle, 11, 13, 14, 16
Permanently deformed nucleus
  angular momentum in, 26
  quantum numbers for, 31
  rotation of, 27
  single-particle excitations in, 29—34
  vibration of, 29
Phonon-octupole excitation, 29
Phonons, 15, 24
Photoelectric cross sections, 84
Photoelectric effect, 83—84
Photon, 80
  absorption of energy of, 88
  energy of, 88
Poisson frequency distribution, 135—139
Positrons, 42
  decay of, 42
  emission of, 62
  ratio of electron capture to emission of, 67—68
Potential
  mean ionization, 97
  square well, 9
Precision index, 127
Principal axis, 24
Probability distribution, 127, 133

Prolate deformation, 24, 26, 30
Protons, 39, 80, 100
  range-energy curves for, 99
  symmetry of, 4, 11

Q

Quadrupole
  electric, 71
  magnetic, 71
  vibrations of, 24
Quadrupole moments, 2, 7, 12, 26
Quanta (phonons), 15, 24
Quantum numbers
  doubly magic nuclides, 18
  neighboring nuclides, 18
  permanently deformed nucleus, 31
Quasiparticle, 23

R

Radiation
  annihilation, 42, 80
  electric multipole, 72, 74
  electromagnetic, 8, 80
  magnetic multipole, 72, 74
  multipole, see Multipole radiation
Radioactive decay, 119—121
  continuum theory of, 107—126
  curve for, 38—39
  fundamental law of, 107
  random nature of, 106
  statistical fluctuations in, 126—159
  systematics of, 39—47
Radioactive disintegration, 38, 39
Radioactive growth, 119—121
Radioactivity
  natural, 38
  units of, 110
Radionuclides, see Nuclides
Random-coincidence rate, 147—148
Random nature of radioactive decay, 106
Range, 82, 101
  extrapolated, 97
  nuclear forces, 9, 11, 12
Range-energy curves, 101
  proton, 99
Relativistic energies, 100
Repulsion, 3
Repulsive core, 9, 11
Rest mass, 41
Root-mean square (rms) deviation, see Standard deviation
Rotating permanently deformed nucleus, 27
Rotational band, 28
Rotational excitations, 27—29
  spin of, 28
Rotational motion of nucleus, 22, 27
Rotation of deformed nucleus, 22, 27
Rutherford (unit), 38

S

Saturation
    nuclear density, 2
    nuclear forces, 9
Scattering
    Compton, 88, 92
    cross section of, 82, 85
    differential, 82, 85
    elastic, 102—103
    inelastic, 103
    photon energy, 88
    Thomson, 86
Schrödinger equation, 11, 15, 18
Secular equilibrium, 117
Shape of nucleus, 2—3, 13
Shell, 14, 16
    closed, 13
    correction for, 96, 97
Single-particle excitations, 16
    parity of, 31
    permanently deformed nucleus and, 29—34
    spin of, 31
Single-particle model of nucleus, 14, 16, 18
Singlet state, 11
Spectra
    alpha particles, 51—52
    beta particles, 59—60
    gamma rays, 68
Spherical harmonics, 23
Spherical nucleus, 13, 15, 18
Spheroidal nucleus, 26
Spin, 4, 28
    doubly magic nuclides, 19
    intrinsic, 4, 16
    neighboring nuclides, 19
    nucleus, 4, 18
    rotational excitations, 28
    single-particle excitations in permanently
        deformed nucleus, 31
    vibrational excitations, 29
Spin exchange, 10
Spin-orbit interaction, 12, 18, 26
Spring constant, 14, 29
Square well potential, 9
Stable end products, 121—122
Standard deviation, 129, 131
    fractional, 132
    F-test for consistency of, 152—153
Statistical fluctuations in radioactive decay,
        126—159
Stopping power, 100
    Bohr formula for, 96
    mass, 100
    mixture, 97
Straggling, 98
Surface effects, 3
Symbolism for energy states of nucleus, 16, 18, 31
Symmetry
    neutron, 4, 11
    proton, 4, 11
Systematics of radioactive decay, 39—47

T

Tau neutrino, 41
Technetium-99m generator, 125—126
Tensor force, 12
Tests
    chi-square, 155—156
    goodness-of-fit, 152—154
Thomson scattering, 86
Total cross section, 86
Transitions, 64
Transparency, 53, 54
Triplet state, 11
T-test for consistency of means, 152
Two means consistency, 155
Two-nucleon force, 8—13

U

Uncharged particles, 80

V

Velocity-dependent forces, 12
Vibrating permanently deformed nucleus, 29
Vibrational excitations, 29
Vibrational motion of nucleus, 22, 23
Vibrations
    deformed nucleus, 22, 23
    dipole, 24
    frequency of, 14
    octupole, 24
    permanently deformed nucleus, 29
    quadrupole, 24
Volume of nucleus, 9

W

Weak coupling, 23, 25
Well potential, 9
Wigner force, 9, 13

X

X-rays, 43
    characteristic, 80
    gamma ray interactions with, 80—94

Z

Zero charge, 63